Jagdish Chand Bansal · Prashant Jamwal ·
Shahid Hussain

Editors

Sustainable Computing and Intelligent Systems

Proceedings of SCIS 2025, Volume 4

 Springer

Editors
Jagdish Chand Bansal
Department of Mathematics
South Asian University
New Delhi, Delhi, India

Prashant Jamwal
Electrical and Computer Engineering
School of Engineering and Digital Sciences
Astana, Kazakhstan

Shahid Hussain
Biomedical Robotics, Information
Technology (IT) and Systems
University of Canberra
Canberra, ACT, Australia

ISSN 2367-3370   ISSN 2367-3389 (electronic)
Lecture Notes in Networks and Systems
ISBN 978-3-032-22910-6   ISBN 978-3-032-22911-3 (eBook)
https://doi.org/10.1007/978-3-032-22911-3

© The Editor(s) (if applicable) and The Author(s) 2026. This book is an open access publication.

Open Access This book is licensed under the terms of the Creative Commons Attribution-NonCommercial-NoDerivatives 4.0 International License (http://creativecommons.org/licenses/by-nc-nd/4.0/), which permits any noncommercial use, sharing, distribution and reproduction in any medium or format, as long as you give appropriate credit to the original author(s) and the source, provide a link to the Creative Commons license and indicate if you modified the licensed material. You do not have permission under this license to share adapted material derived from this book or parts of it.

The images or other third party material in this book are included in the book's Creative Commons license, unless indicated otherwise in a credit line to the material. If material is not included in the book's Creative Commons license and your intended use is not permitted by statutory regulation or exceeds the permitted use, you will need to obtain permission directly from the copyright holder.

This work is subject to copyright. All commercial rights are reserved by the author(s), whether the whole or part of the material is concerned, specifically the rights of translation, reprinting, reuse of illustrations, recitation, broadcasting, reproduction on microfilms or in any other physical way, and transmission or information storage and retrieval, electronic adaptation, computer software, or by similar or dissimilar methodology now known or hereafter developed. Regarding these commercial rights a non-exclusive license has been granted to the publisher.

The use of general descriptive names, registered names, trademarks, service marks, etc. in this publication does not imply, even in the absence of a specific statement, that such names are exempt from the relevant protective laws and regulations and therefore free for general use.

The publisher, the authors and the editors are safe to assume that the advice and information in this book are believed to be true and accurate at the date of publication. Neither the publisher nor the authors or the editors give a warranty, expressed or implied, with respect to the material contained herein or for any errors or omissions that may have been made. The publisher remains neutral with regard to jurisdictional claims in published maps and institutional affiliations.

This Springer imprint is published by the registered company Springer Nature Switzerland AG
The registered company address is: Gewerbestrasse 11, 6330 Cham, Switzerland

If disposing of this product, please recycle the paper.

Lecture Notes in Networks and Systems   1929

Series Editor

Janusz Kacprzyk, *Systems Research Institute, Polish Academy of Sciences, Warsaw, Poland*

Advisory Editors

Fernando Gomide, *Department of Computer Engineering and Automation—DCA, School of Electrical and Computer Engineering—FEEC, University of Campinas—UNICAMP, São Paulo, Brazil*

Okyay Kaynak, *Department of Electrical and Electronics Engineering, Bogazici University, Istanbul, Türkiye*

Derong Liu, *Department of Electrical and Computer Engineering, University of Illinois at Chicago, Chicago, USA*

   Institute of Automation, Chinese Academy of Sciences, Beijing, China

Witold Pedrycz, *Department of Electrical and Computer Engineering, University of Alberta, Edmonton, Alberta, Canada*

   Systems Research Institute, Polish Academy of Sciences, Warsaw, Poland

Marios M. Polycarpou, *Department of Electrical and Computer Engineering, KIOS Research Center for Intelligent Systems and Networks, University of Cyprus, Nicosia, Cyprus*

Imre J. Rudas, *Óbuda University, Budapest, Hungary*

Jun Wang, *Department of Computer Science, City University of Hong Kong, Kowloon, Hong Kong*

The series "Lecture Notes in Networks and Systems" publishes the latest developments in Networks and Systems—quickly, informally and with high quality. Original research reported in proceedings and post-proceedings represents the core of LNNS.

Volumes published in LNNS embrace all aspects and subfields of, as well as new challenges in, Networks and Systems.

The series contains proceedings and edited volumes in systems and networks, spanning the areas of Cyber-Physical Systems, Autonomous Systems, Sensor Networks, Control Systems, Energy Systems, Automotive Systems, Biological Systems, Vehicular Networking and Connected Vehicles, Aerospace Systems, Automation, Manufacturing, Smart Grids, Nonlinear Systems, Power Systems, Robotics, Social Systems, Economic Systems and other. Of particular value to both the contributors and the readership are the short publication timeframe and the worldwide distribution and exposure which enable both a wide and rapid dissemination of research output.

The series covers the theory, applications, and perspectives on the state of the art and future developments relevant to systems and networks, decision making, control, complex processes and related areas, as embedded in the fields of interdisciplinary and applied sciences, engineering, computer science, physics, economics, social, and life sciences, as well as the paradigms and methodologies behind them.

Indexed by SCOPUS, EI Compendex, INSPEC, WTI Frankfurt eG, zbMATH, SCImago.

All books published in the series are submitted for consideration in Web of Science.

For proposals from Asia please contact Aninda Bose (aninda.bose@springer.com).

Preface

This volume presents the fourth collection of selected research papers from the International Conference on Sustainable Computing and Intelligent Systems (SCIS 2025), held on November 7–8, 2025, at the University of Canberra, Australia, under the technical sponsorship of the Soft Computing Research Society, India. Volume 4 brings together 15 rigorously peer-reviewed papers that emphasize theoretical advances, intelligent modeling, and practical AI-driven solutions across diverse application domains.

The contributions in this volume reflect significant progress in machine learning theory, optimization, and intelligent decision systems. Several studies focus on advanced learning paradigms, including preference modeling using OWA operators, echo state networks optimized via metaheuristics, reinforcement learning-driven mutation strategies in differential evolution, and event-driven spiking neural networks. These works contribute to both the interpretability and adaptability of intelligent systems.

Applied AI forms a strong pillar of this volume, with papers addressing healthcare analytics, such as hybrid ensemble models for stroke risk prediction, AI-based virtual interview systems with emotion detection, and scientometric insights into gesture recognition research. Security and trust in digital ecosystems are explored through novel approaches for detecting interest flooding attacks in information-centric networking and polarity-aware fake news detection on social media platforms.

The volume also highlights AI for sustainability and society, including smart irrigation systems for market gardening, pest infestation detection and forecasting, and remote sensing-based land classification. Emerging topics such as personality trait prediction using large language models, text-guided artistic style transfer, and adaptive synthetic image generation further demonstrate the interdisciplinary breadth of SCIS 2025.

We extend our sincere appreciation to the authors, reviewers, and organizing committee for their dedication. We hope this volume serves as a valuable reference and inspires continued innovation in sustainable computing and intelligent systems.

Jagdish Chand Bansal
Prashant Jamwal
Shahid Hussain

SCIS 2025 Organizing Committee

General Chair

Shahid Hussain   University of Canberra, Australia
Prashant Jamwal   Nazarbayev University, Kazakhstan
Jagdish Chand Bansal   South Asian University, New Delhi, India

Program Chair(s)

Vivek Shrivastava   National Institute of Technology, Uttarakhand, India
D L Suthar   Wollo University, Ethiopia
Neeraj Rathore   Indira Gandhi National Tribal University, Amarkantak, Madhya Pradesh, India

Organizing Secretary

Sandeep Kumar   Christ University (Deemed to be University), Bengaluru, Karnataka, India
Satyasai Jagannath Nanda   Malaviya National Institute of Technology Jaipur, Rajasthan, India
Sakshi Shringi   Manipal University Jaipur, Rajasthan, India

Organizing Chair

Harish Sharma   Rajasthan Technical University, Kota, Rajasthan, India
Manish Gupta   Torrens University, Adelaide City Centre, Australia
Saroj Hirenwal   Victorian Institute of Technology, Adelaide Campus, South Australia

Organizing Committee

Prashant Singh Rana	Thapar Institute of Technology, Patiala, Punjab, India
S D Purohit	Rajasthan Technical University, Kota, Rajasthan, India
Neha Yadav	Dr. B. R. Ambedkar National Institute of Technology Jalandhar, Punjab, India
Lokesh Chouhan	National Forensic Sciences University, Goa Campus, India
Kusum Kumari Bharati	Dr. B. R. Ambedkar National Institute of Technology Jalandhar, Punjab, India
Abhishek Verma	Indian Institute of Management (IIM) Rohtak, Haryana, India
Ashish Tripathi	Malaviya National Institute of Technology Jaipur, Rajasthan, India
Faruk Ucar	Marmara University, Istanbul, Türkiye
Himanshu Mittal	Indira Gandhi Delhi Technical University for Women, Delhi, India
K S Nisar	Prince Sattam bin Abdulaziz University, Riyadh, Saudi Arabia

Publicity Chair

Mukesh Saraswat	Jaypee Institute of Information Technology, Noida, Uttar Pradesh, India
Neenu Sheoran	Challenger Limited Sydney, Australia
Avinash Chandra Pandey	PDPM Indian Institute of Information Technology, Design, and Manufacturing, Jabalpur, Madhya Pradesh, India
Anupam Yadav	Dr. B. R. Ambedkar National Institute of Technology Jalandhar, Punjab, India

Technical Program Committee

Muzeeb Mohammad	Georgia Institute of Technology, Atlanta, GA, USA
Narasimha Rao Vajjhala	University of New York Tirana, Tirana, Albania
Jinwei Liu	Florida A&M University, Florida
Riki Patel	Florida Atlantic University, Florida
Vivekananda Jayaram	Florida International University, Florida

Puneet Gangrade	Fordham University, New York, United States
Raju Pal	Gautam Buddha University, Greater Noida, Uttar Pradesh, India
Venkata Ashok K Gorantla	Georgia Institute of Technology, Atlanta, Georgia
Ajay Sharma	Government Engineering College Jhalawar, Rajasthan, India
Ankit Verma	Guru Gobind Singh Indraprastha University, Delhi, India
Rita Yi Man Li	Hong Kong Shue Yan University, Hong Kong
Rajani Kumari	IBS University, Bangalore, India
Sourabh Jain	IIIT Sonepat, Haryana, India
Anshul Verma	IIT-BHU, Uttar Pradesh, India
Jyotindra Narayan	Imperial College London, England
Quang-Vinh Dang	Industrial University of Ho Chi Minh City, Vietnam
Tania Pencheva	Institute of Biophysics and Biomedical Engineering — Bulgarian Academy of Sciences, Bulgaria
Raul Duarte Salgueiral Gomes Campilho	ISEP — School of Engineering, Porto, Portugal
Sonal Sharma	JAIN University, Bengaluru, India
Nupur Jangu	K. R. Mangalam University, Gurugram, India
Elangovan Guruva Reddy	Koneru Lakshmaiah Education Foundation, Vijayawada, AP, India
Joong Hoon Kim	Korea University, South Korea
Philip Moore	Lanzhou University, China
Neil Buckley	Liverpool Hope University UK, England
Vikas Bajpayee	LNMIIT Jaipur, Jaipur Rajasthan India
Dharmendra Jain	Mahaveer Institute of Technology & Science Hyderabad, Telangana, India
Sayar Singh Shekhawat	Manipal University Jaipur, Rajasthan, India
Varun Tiwari	Manipal University, Jaipur, Rajasthan, India
Sweta Jain	Maulana Azad National Institute of Technology, Bhopal, Madhya Pradesh, India
Debangana Ram	Maulana Azad National Institute of Technology, Madhya Pradesh, India
Husam Yahya Abdulwahid	Ministry of Higher Education and Scientific Research, Baghdad, Iraq.
D Sirisha	Nadimpalli Satyanarayana Raju Institute of Technology, Andhra Pradesh, India
Meng-Hiot Lim	Nanyang Technological University, Singapore
Ramakrishnudu Tene	National Institute of Technology Warangal, Telangana, India

Md. Hayder Ali	National University Bangladesh, Bangladesh
Mazin R. Al-Hameed	National University of Science and Technology: Nasiriyah, Dhi Qar, Iraq
Sadouanouan Malo	Nazi BONI University, Burkina Faso
Nirmala M	New Horizon College of Engineering, Bengaluru, Karnataka, India
Ferdib-Al-Islam	Northern University of Business and Technology Khulna, Bangladesh
Suleyman Nokerov	Oguz Han Engineering and Technology University of Turkmenistan, Ashgabat
Manish Dubey	Poornima College of Engineering, Rajasthan, India
K S Nisar	Prince Sattam bin Abdulaziz University, Riyadh, Saudi Arabia
Nahid Fatima	Prince Sultan University Riyadh, Saudi Arabia
Shiv Kant	Raffles University, Rajasthan, India
P. K. Verma	Rajkiya Engineering College Sonbhadra, Uttar Pradesh, India
Shimpi Singh Jadaon	Rajkiya Engineering College, Kannauj, Uttar Pradesh, India
Pramod Sharma	Regional College for Education Research and Technology, Rajasthan, India
Manish Kumar	RV University, Bangalore, India
Chandra Singh	Sahyadri College of Engineering & Management, Mangaluru, India
R.Manikandan	SASTRA University, Tamil Nadu, India
K B Sathya	Sathyabama Institute of Science and Technology, Deemed University, Tamil Nadu, India
Ukasz Marzantowicz	SGH Warsaw School of Economics, Poland
Prasham Sheth	Shri Lal Bahadur Shastri National Sanskrit University, New Delhi, India
Kalyani Nikhilesh Pampatti	SIES College of Arts, Science & Commerce, Mumbai, Maharashtra, India
Seemant Tiwari	Southern Taiwan University of Science and Technology, Tainan City, Taiwan
Anand Upadhyay	St. Francis Institute of Management and Research, Mumbai, Maharashtra, India
Elena Doynikova	St.Petersburg Institute for Informatics and Automation of the Russian Academy of Sciences, Russia
Ruhul amin	Sylhet International University, Shamimabad, Bagbari, Sylhet, Bangladesh
Nupa Ram Chauhan	Teerthanker Mahaveer University, Moradabad, Uttar Pradesh, India

Pavlo Maruschak	Ternopil Ivan Puluj National Technical University, Ternopil, Ukraine
Puneet Kumar Gupta	The ICFAI University Dehradun, India
Man Fung Lo	The University of Hong Kong, Hong Kong
Rahul Ranjeev Kumar	The University of the South Pacific, Suva, Fiji
Libor Peka	Tomas Bata University in Zlin & College of Polytechnics Jihlava, Czechia
Hari K.C.	Tribhuvan University, Institute of Engineering Pashchimanchal Campus, Pokhara, Nepal
Lei Kou	TUD Dresden University of Technology, Germany
Thangavel Murugan	United Arab Emirates University, Al Ain
Antonio Sarasa-Cabezuelo	Universidad Complutense de Madrid, Spain
Reynaldo Ricardo Quispe Infantes	Universidad Nacional Jose Maria Arguedas, Andahuaylas, Peru
Maizate Abderrahim	Universite Hassan II DE Casablansa, Morocco
Rafidah Abd Karim	Universiti Teknologi Mara, Malaysia
Soundes Belkacem	University of Batna 2, Algeria
Madan Kumar Sharma	University of Buraimi, Oman
Dipanwita Thakur	University of Calabria, Italy
Rathish Mohan	University of Cincinnati, Ohio, United States
Beata Dratwiska-Kania	University of Economics in Katowice, Poland
Mazen Ismaeel Ghareb	University of Human Development, Sulimanaya, Iraq
Luiz Guerreiro Lopes	University of Madeira, Portugal
Olugbemiga Solomon Popo La	University of Medical Sciences, Ondo City, Nigeria
Manideep Marripudugala	University of New Haven, Connecticut
Nicholas Eze	University of Nigeria, Nsukka, Nigeria
Ashraful Islam	University of Regina, Saskatchewan, Canada
Amudhavel	University of Tartu, Estonia
Sami Ouali	University of Technology and Applied Sciences, Muscat, Oman
Mohd Shahid Husain	University of Technology and Applied Sciences, Oman
Jiehua Zhou	University of Texas Medical Branch, United States
Santosh Reddy Addula	University of the Cumberlands, Kentucky
Dharyll Prince Abellana	University of the Philippines Cebu, Philippines
Shruti Singh	Washington State University, Washington
Maamri Fouzia	Abbes Laghrour University, Khenchela, Algeria
Rachid Herbazi	Abdelmalek Essaadi University, Tangier, Morocco

Ashish Kumar Mourya	ABES Institute of Technology Ghaziabad, Uttar Pradesh, India
Isak Karabegovi	Academy of Sciences and Arts of Bosnia and Herzegovina, Bistrik, Bosnia and Herzegovina
Kazi A. Kalpoma	Ahsanullah University of Science and Technology (AUST), Dhaka, Bangladesh
Md Afroz	Al Faisal University, Prince Sultan College of Business, Jeddah
Neetu Gupta	Amity University Uttar Pradesh, Noida, India
Sanjay Jain	Amity University, Jaipur, Rajasthan, India

Contents

About the Editors

Dr. Jagdish Chand Bansal is Associate Professor at South Asian University New Delhi and Visiting Faculty at Maths and Computer Science, Liverpool Hope University UK. Dr. Bansal has obtained his Ph.D. in Mathematics from IIT Roorkee. Before joining SAU New Delhi, he has worked as Assistant Professor at ABV-Indian Institute of Information Technology and Management Gwalior and BITS Pilani. His primary area of interest is Swarm Intelligence and Nature-Inspired Optimization Techniques. Recently, he proposed a fission-fusion social structure-based optimization algorithm, Spider Monkey Optimization (SMO), which is being applied to various problems from engineering domain. He has published more than 70 research papers in various international journals/conferences. He is Editor in Chief of the journal MethodsX published by Elsevier. He is Series Editor of the book series Algorithms for Intelligent Systems (AIS) and Studies in Autonomic, Data-driven and Industrial Computing (SADIC) published by Springer. He is Editor in Chief of International Journal of Swarm Intelligence (IJSI) published by Inderscience. He is also Associate Editor of Engineering Applications of Artificial Intelligence (EAAI) and ARRAY published by Elsevier. He is General Secretary of Soft Computing Research Society (SCRS). He has also received Gold Medal at UG and PG level.

Dr. Hussain is working at University of Canberra as Associate Professor of Biomedical Robotics. Prior to that he has worked as Lecturer at University of Wollongong, Australia. Dr. Hussain has obtained his PhD in Mechanical Engineering from the University of Auckland, New Zealand, in 2013. His research interests include assistive and rehabilitation robotics, compliant actuation of robots, robot mechanism design and optimization, nonlinear dynamics and control of robotic systems, human–robot interaction, biomechanical modelling, engineering education and micro electro-mechanical systems (MEMS). Dr. Hussain has published more than 65 papers in the prestigious journals of the field.

Prof. Prashant K. Jamwal earned PhD degree and a post-doctoral fellowship from the University of Auckland, New Zealand. Earlier he had obtained M. Tech. from I.I.T., India, securing first position in all the disciplines and B. Tech. from MNREC, Allahabad, India. Presently, he is working as Professor at the School of Engineering and Design Sciences, Nazarbayev University (NU), Astana, Kazakhstan, and as Adjunct Professor at University of Canberra, Australia. He is actively pursuing research in robotics and artificial intelligence, multi-objective evolutionary optimization, biomedical engineering, and renewable energy. Over the past decade, he has applied his research in the development of medical robots for rehabilitation and surgical applications besides developing improved algorithms for cancer data analytics. Prof. Prashant received three prestigious World Bank Grants and has launched two startup companies and established a Center of

Excellence in Medical Research and Robotics. Lately, he has successfully commercialized an autonomous robotic Gait Rehabilitation System after furnishing all the regulatory requirements, thus making expensive medical robots accessible to small clinics. He has more than 25 years of teaching and research experience and has published more than 100 research articles in reputed international journals/conferences. Prof. Prashant has won many awards such as Jewel of India, Asian Universities Alliance Scholar Award, and Best paper awards, and in 2019 United Nations acknowledged one of his robotics projects in top twenty innovative projects in the world. He is working as Editor for the International journal of bio-mechatronics and bio-robotics and as Reviewer to quite a few international journals and conferences of repute. Prof. Jamwal led many international funded research projects and has so far received research grants worth more than $9 million including prestigious World Bank grants.

The Duality of Preference: Reconciling A Priori and A Posteriori Attitudes in Induced OWA Operator

Bablu Kumar Paswan[ID] and Amar Kishor[(✉)][ID]

PG Department of Mathematics, Magadh University, Bodh Gaya, Bihar 824234, India
amarsdma@gmail.com

Abstract. The Ordered Weighted Averaging (OWA) and its variant, the Induced Ordered Weighted Averaging (IOWA), are two significant aggregation operators that assist decision-makers (DMs) in their decision-making processes. The characteristic of OWA operator is its "orness" metric measures its intrinsic optimism or pessimism level. However, the IOWA operator introduces a significant additional notion. The order of the inputs is not fixed in advance. Instead, it is decided by external elements, known as inducing variables. These variables guide how the inputs are arranged before aggregation stage. This paper shows that a lone, static measure of 'attitudinal character' is not enough for IOWA operators. We therefore introduce two complementary ideas. The first is 'intrinsic attitude', which represents the operator's built-in attitudinal bias, similar to the concept of orness (attitudinal character) in OWA operators. The second is 'realized attitude', which describes the actual behavior observed when the operator is applied to the input values. This realized attitude follows from the interaction between the inducing variables and the input values. This dual understanding helps in creating and understanding IOWA-based decision systems better. It makes them clearer and reduces unintentional biases in decision making.

Keywords: Decision Making · Aggregation Operator · IOWA operator · OWA operator · Orness/Attitudinal Character · Information Fusion

1   Introduction

Decision-making generally happens in a complex environment. This means we often have to think about many factors at the same time. Sometimes, these factors are even conflicting in nature with each other. Take the simple instance of buying a new car. We care about safety. We also think about the price, fuel efficiency, and also about the appearance of the car. All these aspects matter together, not one by one. To handle such situations, DM may use aggregation

© The Author(s) 2026
J. C. Bansal et al. (Eds.): SCIS 2025, LNNS 1929, pp. 1–17, 2026.
https://doi.org/10.1007/978-3-032-22911-3_1

operators. These are mathematical tools that help combine different bits of information together turning many inputs into one clear and reasonable result (output).

Among these tools, one of the most important is the OWA operator, introduced by Yager [21]. What makes the OWA operator special is that it reflects the DM's attitude towards the type of aggregation. It does more than just calculate a simple average. Instead, it shows how a person contemplates while making a decision. For example, an optimist may focus more on the best possible outcomes. A pessimist may pay more attention to the worst cases. A realist may try to balance both views staying in middle of these extremes outcomes.

The OWA operator can represent all of these perspectives by varying the weight vector. It can also handle many other types of decision-making conditions. It does this by carefully assigning weights (weight vectors) after ordering (increasing or decreasing) the inputs. This way, it works like a flexible dial. The dial can be adjusted to match different decision-making styles of DM's choice. At the same time, it keeps strong mathematical properties, which ensure the results (output) remain consistent and trustworthy to the DM. Hence, at the core of every OWA operator is its weight vector (also termed as weighting vector) [21]. This weight vector outlines how the aggregation works. It symbolizes the character, or "personality," of the decision-making process.

Hence, choosing these weight vectors is very important. In fact, it has been a major focus of research for many years. As a result, researchers have established four broad methodologies to determine them as follows:

- **Programming-based techniques:** In this approach, the weight vectors are obtained using optimization (programming) techniques. These methods allow the DM to clearly control the level of optimism (pessimism) in the aggregation process [7,10].
- **Quantifier-based techniques:** This approach starts with natural language expressions such as "most" or "about half." These linguistic terms are then converted into numerical weights (weight vectors). In this way, everyday language is linked with mathematical modeling [3,24,26].
- **Experience-based techniques:** Here, weights are chosen based on practical experience of the DM or expert assisting the DM. Expert knowledge and judgment play a key role in this process. This helps anchor the OWA operator in real-world decision-making situations [4,6,17,28].
- **Weight function-based techniques:** In this case, weight vectors are generated directly using predefined mathematical functions (formula). These functions provide a clear and systematic way to obtain the weight vectors. As a result, the procedure becomes easy for replication and validation [5,11,15,19].

The use of OWA operators has developed extensively over time. This growth is largely due to the development of different methodological approaches shown above. Among them, programming-based techniques stand out [10]. They are especially useful because they allow weights to be adjusted precisely to match a desired level of orness [7]. To meet the needs of different applications, many other families of OWA operators have been proposed in the literature. Each family modifies the basic OWA idea in a particular manner. Some well-known examples comprise the Centered OWAO [25] and the Hurwicz OWA (H-OWAO) [22]. Other significant variants are the BK OWAO [18] and the Slide-based OWAO (S-OWAO) family [23,27]. The Olympic OWAO [24] and the Ahn OWA family [2] further expand this framework considerably.

Additional extensions include the Binomial OWA (Bin-OWAO) [9] and the Stancu OWA (Sta-OWAO) [13] based on Stancu distribution. More recent contributions include the Inverse Hypergeometric OWA (In-Hyp OWAO) [8], the Beta–Bezier OWA (BB-OWAO) [15], and the B-OWAO families [14] based on different probability distributions. The Beta–Binomial OWA (BBN-OWAO) [16] represents another important advancement with generalized features and applications. Together, these advances highlight the flexibility and applicability of the OWA framework. They also show its continued evolution in handling complex and realistic decision-making problems. In the classical OWA structure, there is a fundamental drawback in that it ranks its inputs by the numeric values, ignoring other aspects of information. For example, consider an investment decision where the expected return is very high and the estimate comes from an unreliable source. Now consider another reasonably opposing case. The expected return is lower, but it is provided by a highly trusted expert. If we look only at the return values, both cases are treated in the same way. This raises an important query. Should the aggregation depend only on numerical magnitude? In many real-world decisions, the answer is negative. The context or quality of the information also matters in such situations. This context may represent reliability, trustworthiness, and relevance. Often, these factors are just as important as the value itself. To address this particular limitation, Yager and Filev introduced the *Induced Ordered Weighted Averaging (IOWA)* operator [28]. The basic idea behind IOWA is simple but prevailing. In the IOWA framework, each input value x_i is paired with an additional variable. This particular variable is called the *order-inducing variable* and is denoted by h_i. Together, they form the pair $\langle h_i, x_i \rangle$. Unlike classical OWA, the ordering is not based on the numerical values of x_i. Instead, the pairs are reordered (increasing or decreasing) according to the numerical values of h_i. After this reordering, the usual OWA weight vector is applied to the corresponding x_i values. This modification allows the aggregation process to depend on external information. The ordering can now reflect factors other than numerical magnitude. As a result, IOWA becomes well suited for situations involving heterogeneous sources of information [1]. The usefulness of IOWA can be seen in many applications. In social network analysis, for example, expert trust values (h_i) can determine how opinions (x_i) are weighted [12]. Similarly, in multi-criteria decision-making problems, criteria importance or reliability can guide the aggregation order [20]. Although IOWA operators have been widely studied and applied, one important matter has not received adequate attention. This issue concerns the true meaning of the operator's "attitudinal character (orness)." In the case of classical OWA operators, this concept is well understood. Yager's measure of "orness" [21] provides a single numerical value. This value clearly indicates whether an operator behaves in an optimistic or pessimistic manner or in between. Importantly, this measure depends only on the weight vector. It remains fixed, regardless of the input data (values). The situation is altered for IOWA operators. Here, the ordering of inputs is not fixed. Instead, it changes dynamically according to the inducing variables applied. This naturally leads to a key question. Does the orness computed from the IOWA operator's weight vector accurately describe how the operator truly behaves? Or does the combined effect of inducing variables and input values produce a new behavior altogether? In many cases, this resulting behavior may deviate significantly from what the operator was originally designed to express.

To address this problem, this paper presents a new perspective based on discrepancy in the attitudinal character (orness) of IOWA operators. In short, the core argument is that a single numerical measure is insufficient to fully describe the operator's true attitude [4]. Instead, we show that an IOWA operator possesses two closely connected, yet clearly different, attitudinal components, which are eloberated in the following sections.

- **Intrinsic Disposition:** This is a static quantity determined by the IOWA weigh vector alone. It represents the desired attitude of the operator applied. More specifically, it is how the operator intends to weigh the inputs after the inputs have been ordered according to the inducing variables. This component is exactly like the standard orness of the OWA operators [21].
- **Realized Attitude:** This characterizes a new computation that serves as a complement to the intrinsic disposition discussed earlier. It is based upon the actual data to be aggregated and Influenced by the actual values of $\mathbf{h}$ and $\mathbf{x}$. It tries to measure how the operator performs, in practicality.

The proposed work define the two attitudinal aspects of the IOWA operator, analyzes and studies their relationship, and shows how they behave in practical scenarios. The work also support these findings with suitable examples and simulations, providing a framework that explains the nuanced behavior of IOWA and IOWA-based operators [28]. The paper is organized as follows. Section 2 introduces the basic concepts. In Sect. 3, we present our perspective on intrinsic disposition and realized attitude. Sections 4 and 5 then illustrate how these two attitudes can diverge through examples and simulation studies. Section 6 discusses the broader implications and limitations of the proposed approach, and Sect. 7 concludes the paper.

2   Foundational Concepts: OWA, IOWA, and Orness

In this section, we introduce the main concepts, such as the standard OWA operator and associated terminologies. We then describe the standard IOWA operator and its working. Finally, we revisit the notion of orness or attitudinal character, which plays a central role in our proposed work.

2.1 The Ordered Weighted Averaging (OWA) Operator

The Ordered Weighted Averaging (OWA) operator was originally proposed by R. R. Yager [21]. It provides a structured way to combine a collection of inputs, denoted by $\mathbf{x} = (x_1, \ldots, x_n)$. The aggregation process is guided by a specific weight vector $\mathbf{W} = (w_1, \ldots, w_n)$, where each weight satisfies $0 \leq w_j \leq 1$ and the weights collectively sum to one, that is, $\sum_{j=1}^{n} w_j = 1$.

What differentiates the OWA operator from other aggregation operators is the manner in which the weights are used. The input values are not united in their original order. Instead, they are first arranged from the largest to the smallest. Once this ordering is established, the weights of the weight vector are applied to these ranked values rather than to the original inputs. Let $x_{(j)}$ denote the j-th largest element of the vector $\mathbf{x}$, so that the ordered values satisfy

$$x_{(1)} \geq x_{(2)} \geq \cdots \geq x_{(n)}.$$

Based on this ordered structure, the OWA operator is defined as below [21]:

$$\text{OWA}(\mathbf{x}) = \sum_{j=1}^{n} w_j x_{(j)}. \tag{1}$$

With an appropriate selection of weights, this formulation enables the OWA operator to replicate the behavior of several familiar aggregation operators. In particular, it can act like the minimum (Min), the maximum (Max), or the arithmetic mean, among other aggregation operators.

- **Max OWA Operator:** If $\mathbf{W} = (1, 0, \ldots, 0)$, only the largest value is selected. In this case, $\text{OWA}(\mathbf{x}) = x_{(1)} = \max(\mathbf{x})$. This represents an extremely optimistic attitude.
- **Min OWA Operator:** If $\mathbf{W} = (0, \ldots, 0, 1)$, only the smallest value is considered. Here, $\text{OWA}(\mathbf{x}) = x_{(n)} = \min(\mathbf{x})$. This corresponds to an extremely pessimistic attitude.
- **Arithmetic Mean:** If all weights are equal, i.e., $\mathbf{W} = \left(\frac{1}{n}, \ldots, \frac{1}{n}\right)$, the OWA operator reduces to the arithmetic mean operator. In this case, $\text{OWA}(\mathbf{x}) = \frac{1}{n} \sum_{i=1}^{n} x_i$. This reflects a neutral or balanced attitude.

2.2 The "Orness" of OWA Operators

A vital concept associated with OWA operators is called the "orness." This concept was introduced by Yager [21] along with the definition of OWA operator. Orness labels the attitude of an OWA operator. More explicitly, it indicates how optimistic the aggregation process is. Orness measures how close the operator is to two extreme cases. At one extreme is the logical OR, or the Max OWA operator. This corresponds to an orness value of 1. At the other extreme is the logical AND, or the Min OWA operator which corresponds to an orness value of 0. The orness of an OWA operator be determined by only on its weight vector $\mathbf{W}$. It is defined as follows:

$$\text{orness}(\mathbf{W}) = \sum_{j=1}^{n} \frac{(n-j)}{n-1} w_j. \tag{2}$$

The value of orness constantly lies between 0 and 1. A value closer to 1 indicates a more optimistic behavior. In this case, higher input values are given more importance.

On the contrary, a value closer to 0 indicates a more pessimistic behavior. Here, lower input values dominate the aggregation or decision. An orness value of 0.5 usually represents a neutral or balanced attitude. The arithmetic mean is a common or prevalent example of this case. Overall, the orness provides a simple and intuitive way to understand the inherent bias of an OWA operator or aggregation. While orness captures optimistic behavior, pessimism is described using the concept of *anti-orness*, also known as *andness*. It is defined as $1-$orness and indicates how closely an aggregation operator behaves like the minimum operator.

2.3   The Induced Ordered Weighted Averaging (IOWA) Operator

The Induced Ordered Weighted Averaging (IOWA) operator was proposed by Yager and Filev [28]. It extends the classical OWA operator by introducing an external criterion for ordering the inputs. In the IOWA framework, the input values $\mathbf{x} = (x_1, \ldots, x_n)$ are not ordered directly. Instead, a distinct set of variables is used for this purpose. These variables are called *order-inducing variables* and are denoted by $\mathbf{h} = (h_1, \ldots, h_n)$. Each input value x_i is paired with its corresponding inducing variable h_i. This forms a collection of pairs of the form $\langle h_i, x_i \rangle$. The aggregation process in IOWA takes place in two main steps:

1. First, the pairs $\langle h_i, x_i \rangle$ are reordered according to the values of the inducing variables h_i. The ordering is done in descending manner. Let σ be a permutation such that $h_{\sigma(1)} \geq h_{\sigma(2)} \geq \cdots \geq h_{\sigma(n)}$.
2. Next, the OWA weight vector $\mathbf{W} = (w_1, \ldots, w_n)$ is applied. These weights are assigned to the x values that correspond to the reordered inducing variables.

Through this mechanism, IOWA allows the aggregation order to be guided by external information (factor). This makes it possible to reflect factors such as reliability, importance, or relevance in the decision process. Formally, the IOWA aggregation is defined as:

$$\text{IOWA}(\langle h_1, x_1 \rangle, \ldots, \langle h_n, x_n \rangle) = \sum_{j=1}^{n} w_j x_{\sigma(j)}. \tag{3}$$

This mechanism is particularly valuable when the "quality" or "context" of an argument (e.g., reliability, importance, confidence) should dictate its influence in the aggregation, rather than its raw magnitude. For example, if h_i represents the reliability of an expert opinion x_i, then the IOWA operator allows more reliable opinions to receive higher weights, regardless of whether they are high or low in value.

2.4   IOWA and the Challenge to Static Attitudinal Measures

While the IOWA operator clearly generalizes OWA, it complicates the interpretation of "orness". If we simply apply the OWA orness formula (Eq. 2) to the IOWA weighting vector $\mathbf{W}$, we obtain a measure of the operator's intended preference for high-ranked inducing variables. However, this does not necessarily tell us how the operator behaves with respect to the argument values $\mathbf{x}$.

Consider an IOWA operator with a highly optimistic $\mathbf{W}$ (orness ≈ 1). This means it's designed to give high weight to arguments associated with the highest h_i. But what if the x_i values associated with the highest h_i are actually low? In such situations, an operator that is designed to be optimistic may act otherwise in practice. Even though its design favors higher values, the final aggregated result may appear neutral or even pessimistic. This mismatch highlights a key issue. There can be a gap between the intended attitude of the operator and its actual behavior. Recognizing this gap is the key motivation behind our proposed dual perspective framework.

3   The Dual Perspective: Intrinsic Disposition and Realized Attitude

The main argument of this paper is straightforward based on the previous discussions. The attitudinal character of an IOWA operator cannot be fully described by a single, fixed measure. An IOWA operator has an intended scheme. It also shows a particular behavior when applied to real inputs (data). These two aspects are not always the same. To capture this difference, we propose a dual framework. This framework separates the operator's designed strategy from its actual, data-driven behavior.

3.1   Intrinsic Disposition: The Operator's Designed Bias

The intrinsic disposition of an IOWA operator pronounces its built-in attitude. This attitude is determined only by the weight vector $\mathbf{W}$ used for aggregation. It represents the bias that is intentionally designed into the operator. This concept is directly comparable to the classical orness measure of an OWA operator. Both measures describe how the operator is meant to behave, before any data or input is applied. Let $\mathbf{W} = (w_1, \ldots, w_n)$ be the weight vector of an IOWA operator. Each weight satisfies $w_j \in [0, 1]$. All weights sum to one, i.e., $\sum_{j=1}^{n} w_j = 1$. The intrinsic disposition is denoted by $\mathcal{D}_{\text{int}}(\mathbf{W})$. It is defined as follows:

$$\mathcal{D}_{\text{int}}(\mathbf{W}) = \sum_{j=1}^{n} \frac{(n-j)}{n-1} w_j. \tag{4}$$

This expression is mathematically identical to the orness measure given in Eq. 2. It indicates how strongly the operator lay emphasis on higher-ranked inducing variables. The interpretation of this measure is intuitive:

- If $\mathcal{D}_{\text{int}}(\mathbf{W})$ is close to 1, the operator is intrinsically optimistic. In this case, arguments associated with the largest h_i values receive more prominence.
- If $\mathcal{D}_{\text{int}}(\mathbf{W})$ is close to 0, the operator is intrinsically pessimistic. Here, arguments paired with smaller h_i values dominate the aggregation.
- If $\mathcal{D}_{\text{int}}(\mathbf{W})$ is close to 0.5, the operator exhibits a neutral or balanced attitude. No particular inducing rank is strongly favored.

Most importantly, $\mathcal{D}_{\text{int}}(\mathbf{W})$ is a static property. It depends only on the operator itself. It does not change with different input data $(\mathbf{h}, \mathbf{x})$. In this sense, the intrinsic disposition represents the policy or design philosophy of the IOWA operator.

3.2   Realized Attitude: The Emergent Behavior in Practice

The realized attitude of an IOWA operator captures its actual attitudinal behavior with respect to the argument values $\mathbf{x}$, given a specific dataset $(\mathbf{h}, \mathbf{x})$. This measure quantifies the effective optimism or pessimism once the order-inducing variables have reshaped the aggregation.

To define realized attitude, we first need to understand which weights from $\mathbf{W}$ are effectively applied to the sorted argument values $\mathbf{x}$. Let:

- σ be the permutation that orders the inducing variables $\mathbf{h}$ in descending order: $h_{\sigma(1)} \geq h_{\sigma(2)} \geq \cdots \geq h_{\sigma(n)}$.
- π be the permutation that orders the argument values $\mathbf{x}$ in descending order: $x_{\pi(1)} \geq x_{\pi(2)} \geq \cdots \geq x_{\pi(n)}$.

The IOWA operator applies weight w_j to $x_{\sigma(j)}$. To determine the realized attitude, we need to find which original weight w_k is effectively assigned to $x_{\pi(j)}$ (the j-th largest argument value).

We construct an effective weighting vector $\mathbf{W}_{\text{eff}} = (w_1', \ldots, w_n')$, where each w_j' is the weight from $\mathbf{W}$ that is applied to the j-th largest argument value $x_{\pi(j)}$. Specifically, w_j' is the weight w_k such that $x_{\pi(j)} = x_{\sigma(k)}$. This implies that $k = \sigma^{-1}(\pi(j))$, where σ^{-1} is the inverse permutation of σ. So, the effective weighting vector elements are:

$$w_j' = w_{\sigma^{-1}(\pi(j))}, \quad j = 1, \ldots, n. \tag{5}$$

Once $\mathbf{W}_{\text{eff}}$ is constructed, the realized attitude, denoted $\mathcal{A}_{\text{real}}(\mathbf{W}, \mathbf{h}, \mathbf{x})$, is defined using the standard orness formula on this effective weighting vector:

$$\mathcal{A}_{\text{real}}(\mathbf{W}, \mathbf{h}, \mathbf{x}) = \sum_{j=1}^{n} \frac{(n-j)}{n-1} w_j'. \tag{6}$$

The realized attitude value, also in $[0, 1]$, quantifies the effective optimism or pessimism of the aggregation with respect to the actual argument values $\mathbf{x}$ for the given input. It is a dynamic, data-dependent measure.

3.3 Intuitive Analogy: Intent Vs. Outcome

Imagine a university admissions committee (the IOWA operator) that has an intrinsic policy (intrinsic disposition) to heavily favor applicants with strong recommendation letters (h_i). This policy is designed to be highly "optimistic" about well-supported candidates. However, in a particular year, the strongest recommendations might come from applicants with only average academic scores (x_i). In this scenario, the committee's "realized attitude" towards academic excellence might appear quite neutral or even pessimistic, despite its strong intrinsic disposition. The final outcome depends on two things. One aspect mirrors the committee's underlying policy, while the other depends on the particular set of applicants. The final outcome is the result from the interaction between these two factors.

3.4 Theoretical Implications of the Discrepancy Observed

The difference between intrinsic disposition and realized attitude is not just terminological; it has meaningful theoretical and practical consequences described below:

- **Interpretability:** Distinguishing the two concepts enhances the transparency of IOWA operators. It clear up their intended design, while revealing how they actually behave on explicit datasets.
- **Insight:** A gap between $\mathcal{D}_{\text{int}}$ and $\mathcal{A}_{\text{real}}$ offers valuable investigative informations. It may signal a misalignment between the inducing variables **h** and the input values **x**. This suggests that the inducing structure does not fully capture the decision perspective.
- **Robustness:** Examining the range of $\mathcal{A}_{\text{real}}$ for a fixed $\mathcal{D}_{\text{int}}$ helps assess the sensitivity of the operator to data deviations. In particular when the relationship between **h** and **x** is uncertain.

4 Illustrative Example

To clarify the ideas of intrinsic disposition and realized attitude, we contemplate a simple example where five contending projects are evaluated by two criteria: their expected impact and their level of environmental compliance.

Description of the Scenario. Consider a funding agency that has to assess five project proposals. Each project i is described by two quantitative attributes below:

- x_i: where x_i represents the expected project impact, measured on a scale from 0 to 100;
- h_i: where h_i is the environmental compliance level, going from 0 (no compliance) to 1 (full compliance).

The agency experts aggregate these assessments using an IOWA operator, directed by a weight vector $\mathbf{W}$ that strongly prioritizes projects with high environmental compliance.

The project data are:

Project ID i	h_i	x_i
1	0.75	60
2	0.90	40
3	0.20	80
4	0.50	30
5	0.85	90

The funding agency's weighting vector, designed to be quite optimistic towards higher-ranked compliance scores, is:

$$\mathbf{W} = (w_1, w_2, w_3, w_4, w_5) = (0.60,\ 0.25,\ 0.10,\ 0.05,\ 0.00).$$

Step 1: Calculate the Intrinsic Disposition. The intrinsic disposition, calculated as $\mathcal{D}_{\text{int}}(\mathbf{W})$, reflects the inherent bias of the agency's policy. Using Eq. 4:

$$\mathcal{D}_{\text{int}}(\mathbf{W}) = \sum_{j=1}^{5} \frac{(5-j)}{4} w_j$$

$$= \frac{1}{4}\left[(4) \cdot 0.60 + (3) \cdot 0.25 + (2) \cdot 0.10 + (1) \cdot 0.05 + (0) \cdot 0.00\right]$$

$$= \frac{1}{4}[2.40 + 0.75 + 0.20 + 0.05 + 0]$$

$$= \frac{3.40}{4} = \mathbf{0.85}.$$

The intrinsic disposition value of 0.85, indicates a strongly optimistic stance regarding environmental compliance. Hence, the agency intends to prioritize projects with high compliance scores heavily.

Step 2: Determine Ordering Permutations.

Ordering by Environmental Compliance (h). Sort the h_i values in descending order and identify the corresponding original project IDs:

$$h = (0.75, 0.90, 0.20, 0.50, 0.85) \xrightarrow{\text{sorted desc}} (0.90, 0.85, 0.75, 0.50, 0.20).$$

The permutation σ that maps ranks to original indices is: $\sigma = (2, 5, 1, 4, 3)$. This means $h_{\sigma(1)} = h_2 = 0.90$ (rank 1), $h_{\sigma(2)} = h_5 = 0.85$ (rank 2), etc. The ordered impact scores for IOWA aggregation are $x_{\sigma(j)} = (x_2, x_5, x_1, x_4, x_3) = (40, 90, 60, 30, 80)$.

Ordering by Impact Score ($\mathbf{x}$). Sort the x_i values in descending order and identify the corresponding original project IDs:

$$x = (60, 40, 80, 30, 90) \xrightarrow{\text{sorted desc}} (90, 80, 60, 40, 30).$$

The permutation π that maps ranks to original indices is: $\pi = (5, 3, 1, 2, 4)$. This means $x_{\pi(1)} = x_5 = 90$ (rank 1), $x_{\pi(2)} = x_3 = 80$ (rank 2), etc.

Step 3: Construct the Effective Weighting Vector ($\mathbf{W}_{\text{eff}}$). The effective weighting vector $\mathbf{W}_{\text{eff}} = (w_1', \ldots, w_n')$ applies weights from $\mathbf{W}$ to the sorted impact scores $\mathbf{x}_{\pi(j)}$. We need to find $w_j' = w_{\sigma^{-1}(\pi(j))}$. First, we determine the inverse permutation σ^{-1}, which maps an original index to its rank based on $\mathbf{h}$: $\sigma^{-1}(1) = 3$, $\sigma^{-1}(2) = 1$, $\sigma^{-1}(3) = 5$, $\sigma^{-1}(4) = 4$, $\sigma^{-1}(5) = 2$.

Now we compute each w_j':

- $j = 1$: affects $x_{\pi(1)} = x_5 = 90$. Original index is 5. Its rank in $\mathbf{h}$ is $\sigma^{-1}(5) = 2$. So, $w_1' = w_2 = 0.25$.
- $j = 2$: affects $x_{\pi(2)} = x_3 = 80$. Original index is 3. Its rank in $\mathbf{h}$ is $\sigma^{-1}(3) = 5$. So, $w_2' = w_5 = 0.00$.
- $j = 3$: affects $x_{\pi(3)} = x_1 = 60$. Original index is 1. Its rank in $\mathbf{h}$ is $\sigma^{-1}(1) = 3$. So, $w_3' = w_3 = 0.10$.
- $j = 4$: affects $x_{\pi(4)} = x_2 = 40$. Original index is 2. Its rank in $\mathbf{h}$ is $\sigma^{-1}(2) = 1$. So, $w_4' = w_1 = 0.60$.
- $j = 5$: affects $x_{\pi(5)} = x_4 = 30$. Original index is 4. Its rank in $\mathbf{h}$ is $\sigma^{-1}(4) = 4$. So, $w_5' = w_4 = 0.05$.

Thus, the effective weighting vector is:

$$\mathbf{W}_{\text{eff}} = (0.25,\ 0.00,\ 0.10,\ 0.60,\ 0.05).$$

Step 4: Calculate the Realized Attitude. Now, compute the realized attitude $\mathcal{A}_{\text{real}}$ using $\mathbf{W}_{\text{eff}}$ and Eq. 6:

$$\begin{aligned}
\mathcal{A}_{\text{real}}(\mathbf{W}, \mathbf{h}, \mathbf{x}) &= \sum_{j=1}^{5} \frac{(5-j)}{4} w_j' \\
&= \frac{1}{4}\big[(4) \cdot 0.25 + (3) \cdot 0.00 + (2) \cdot 0.10 + (1) \cdot 0.60 + (0) \cdot 0.05\big] \\
&= \frac{1}{4}\big[1.00 + 0.00 + 0.20 + 0.60 + 0\big] \\
&= \frac{1.80}{4} = \mathbf{0.45}.
\end{aligned}$$

The realized attitude for this specific dataset is 0.45.

Interpretation of the Discrepancy.

- **Intrinsic Disposition: 0.85 (Highly Optimistic):** The funding agency intended to heavily prioritize projects with good environmental compliance.

- **Realized Attitude: 0.45 (Neutral to Slightly Pessimistic)**: However, when applied to the actual project data, the aggregation ended up behaving in a much more neutral, even slightly pessimistic, manner with respect to the impact scores.

Why does this large difference occur? The main reason is the negative correlation between h_i and x_i. Projects with high environmental compliance tend to have low impact scores. Project 2 is a clear example of this case. On the other hand, projects with high impact scores show poor compliance. Project 3 illustrates this situation.

This discrepancy forces the operator to act in a way that was not originally intended. Weight vector that was chosen to reflect an optimistic one are instead assigned to projects with the lower impact values. On the other hand, projects with high impact scores may be pushed to lower ranks due to weak compliance. This results in smaller effective weights and a key insight: the context introduced by the inducing variables can conclusively shape the final aggregation outcome and substantially alter the apparent attitude of an IOWA operator. Therefore, an operator intentionally designed to be optimistic may show pessimistic behavior. Identifying and understanding this mismatch is crucial for enhancing both the limpidity and effectiveness of IOWA-based decision-making processes.

5   Role of Correlation

The previous example highpoints an important insight: when the inducing variables and input values are not fairly aligned, the operator's behavior can change strikingly. Even a single instance of such misalignment may lead to a noticeable gap. This gap is between the intrinsic disposition and the realized attitude. To study this effect more systematically, we move past a single illustrative case. We conduct a systematized simulation in which the correlation between $\mathbf{h}$ and $\mathbf{x}$ is deliberately changed to observe the effects. This approach allows to clearly assess how different patterns influence the realized behavior of the IOWA operator.

5.1   Simulation Design

We consider a simulation with $n = 10$ inputs and adopt an IOWA weight vector as

$$\mathbf{W} = (0.30,\ 0.25,\ 0.15,\ 0.10,\ 0.08,\ 0.06,\ 0.04,\ 0.02,\ 0.00,\ 0.00),$$

that reflects a moderately optimistic aggregation strategy. For this choice, the intrinsic disposition is $\mathcal{D}_{\mathrm{int}}(\mathbf{W}) \approx 0.794$, indicating an intentional importance on higher-ranked values.

To analyze how dependency affects operator behavior, we generate synthetic pairs of $(\mathbf{h}, \mathbf{x})$ from a bi-variate normal distribution, while systematically varying the correlation coefficient as follows:

$$\rho \in \{-1.0, -0.75, -0.5, -0.25, 0.0, 0.25, 0.5, 0.75, 1.0\}.$$

For each ρ, 10^4 independent trials were performed. In each trial, a dataset of size $n = 10$ generated, the realized attitude $\mathcal{A}_{\text{real}}$ computed, and results have been averaged over all trials. This setup offers a clear view of how correlation between $\mathbf{h}$ and $\mathbf{x}$ shapes the realized behavior of the IOWA operator.

5.2 Simulation Results

Simulation results shown in the Table 1 clearly disclose a steady and methodical relationship between correlation coefficient ρ and the average realized attitude (ARA).

Table 1. Correlation vs. (ARA) Average Realized Attitude

ρ	ARA	Std. Dev.
−1.00	0.206	0.000
−0.75	0.352	0.061
−0.50	0.441	0.083
−0.25	0.495	0.095
0.00	0.548	0.101
0.25	0.601	0.096
0.50	0.658	0.085
0.75	0.724	0.063
1.00	0.794	0.000

The simulation study highlights several key findings, which can be summarized as follows:

1. **Perfect Positive -Correlation ($\rho = 1.0$):** When $\mathbf{h}$ and $\mathbf{x}$ are perfectly aligned, their rankings match exactly. In this case, the realized attitude equals the intrinsic disposition: $\mathcal{A}_{\text{real}} = \mathcal{D}_{\text{int}} \approx 0.794$. The operator's intended optimistic behavior is fully realized.
2. **Perfect Negative -Correlation ($\rho = -1.0$):** When the rankings are perfectly opposite, the realized attitude flips. It becomes approximately the complement of the intrinsic disposition: $\mathcal{A}_{\text{real}} \approx 1 - \mathcal{D}_{\text{int}} = 0.206$. Here, the operator behaves in the exact contradictory way of its design.
3. **No Correlation ($\rho = 0.0$):** When there is no correlation, the realized attitude moves toward a neutral value. It is around 0.55, slightly above 0.5 due to the specific weight distribution. In this case, the operator's designed optimism is largely diluted.
4. **Intermediate Correlations:** For correlations between these extremes, the realized attitude changes gradually. It smoothly interpolates between the pessimistic and optimistic extremes.

5.3   Visualizing the Discrepancy

Figure 1 provides a visual representation of the results from Table 1. The plot clearly illustrates how the realized attitude depends on the correlation between **h** and **x**.

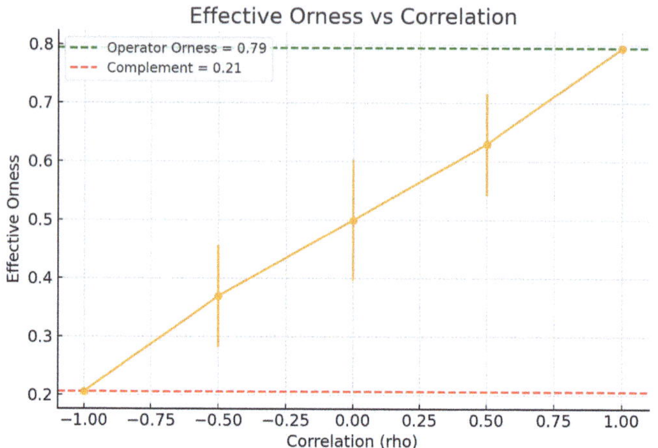

Fig. 1. ARA as a function of ρ, **h**, **x**

This analysis provides a key understanding that the difference between intrinsic disposition and realized attitude is neither an isolated occurrence nor a simple inconsistency. In fact it is a usual feature of IOWA operators in general. Additionally, this discrepancy is predictable as it mainly depends on the degree of association between the inducing variables and input values. When these two values are closely affiliated, the realized attitude closely follows the intrinsic disposition. On the other hand, when there is misalignment, the realized attitude can deviate noticeably.

6   Discussion on the Results

For the clear understanding of the IOWA operators the proposed dual perspective on the attitudinal character—distinguishing between the intrinsic disposition and the realized attitude can be helpful. This perspective identifies the behavior of induced aggregation as dynamic and context-dependent. Separating the operator's intended design from its actual performance, this framework reveals how the IOWA operator reacts to diverse scenarios. Therefore, it propositions a more realistic and nuanced view of how IOWA operators function in practice.

Implications of the Proposed Work The benefits of the proposed work is evident when put into real situations described point wise as follows:

- Presenting both intrinsic disposition and realized attitude leads to greater clarity in model's interpretation and hence its application.
- A noticeable difference here between these two measures should be treated as a warning pointer.
- DMs can examine different correlation patterns between $\mathbf{h}$ and $\mathbf{x}$ during the system design phase. This enables DMs to investigate the range of potential realized attitudes and evaluate the robustness of the decision-making process under the diverse situations.

Limitations and the Future Scope Besides its contributions, this study also points to several restrictions and opportunities for future investigation described below:

- The present study focuses solely at linear correlation. Future work can probe and explore much complex and non-linear relationships between $\mathbf{h}$ and $\mathbf{x}$.
- A natural extension of this work may consider the building of adaptive IOWA operators.
- Another promising path could be the formulation of a unified, normalized metric that can quantify the numerical differences between the $\mathcal{D}_{\mathrm{int}}$ and the $\mathcal{A}_{\mathrm{real}}$. This measure could serve as an easy analytical tool for estimating and comparing aggregation models in decision-support systems in general.
- The distinction observed above, between intended and observed behavior may not be exclusive to IOWA operators. Investigating this broadly, could extend the reach of the proposed perspective across multiple domains and multiple aggregation operators.

7   Conclusion

The proposed work looks closely at problems tied to using just one fixed value to describe how the IOWA operator behaves. A new approach uses two distinct ideas instead. One part captures what the operator is built to do—its inner inclination. The other shows how it actually acts once real inputs come in - the lived stance. These two were named, explained, and studied in depth. Tests with numbers and simulations helped reveal their traits. Often, there's a big difference between expected and actual response. That gap depends heavily on how linked the sorting cues are to the data itself. What emerges: a tool meant to lean hopeful might act flat—or even glum—depending on input patterns. Seeing this split matters greatly when applying these tools wisely. Making this hidden shift visible helps build models people can better follow and trust in terms of aggregation.

References

1. Aggarwal, M.: A new family of induced owa operators. Int. J. Intell. Syst. **30**(2), 170–205 (2015)
2. Seok Ahn, B.: On the properties of OWA operator weights functions with constant level of Orness. IEEE Trans. Fuzzy Syst. **14**(4), 511–515 (2006)
3. Seok Ahn, B.: Some quantifier functions from weighting functions with constant value of orness. IEEE Trans. Syst. Man Cybern. Part B (Cybern.**38**(2), 540–546 (2008)
4. Beliakov, G.: How to build aggregation operators from data. Int. J. Intell. Syst. **18**(8), 903–923 (2003)
5. Calvo, T., Mayor, G., Mesiar, R.: Aggregation operators: new trends and applications, vol. 97. Physica (2012)
6. Filev, D., Yager, R.R.: On the issue of obtaining owa operator weights. Fuzzy Sets Syst. **94**(2), 157–169 (1998)
7. Fullér, R.: On obtaining OWA operator weights: a sort survey of recent developments. In: 2007 IEEE International Conference on Computational Cybernetics, pp. 241–244. IEEE (2007)
8. Kishor, A., Singh, A.K., Sonam, S., Pal, N.R.: A new family of OWA operators featuring constant orness. IEEE Trans. Fuzzy Syst. (2019)
9. León, T., Zuccarello, P., Ayala, G., de Ves, E., Domingo, J.: Applying logistic regression to relevance feedback in image retrieval systems. Patt. Recogn. **40**(10), 2621–2632 (2007)
10. Liu, X.: On the properties of equidifferent OWA operator. Int. J. Approximate Reasoning **43**(1), 90–107 (2006)
11. Liu, X.: A general model of parameterized OWA aggregation with given orness level. Int. J. Approximate Reason. **48**(2), 598–627 (2008)
12. Pérez, L.G., Mata, F., Chiclana, F.: Social network decision making with linguistic trustworthiness-based induced owa operators. Int. J. Intell. Syst. **29**(12), 1117–1137 (2014)
13. Singh, A.K., Kishor, A., Pal., N.R.: Stancu OWA operator. IEEE Trans. Fuzzy Syst. **23**(4), 1306–1313 (2014)
14. Srivastava, V., Kishor, A., Singh, A.K.: Novel optimistic and pessimistic family of OWA operator with constant orness. Int. J. Approximate Reason. **161**, 109006 (2023)
15. Srivastava, V., Singh, A.K.: Beta-bézier owa operator. Int. J. Approximate Reason. **152**, 33–45 (2023)
16. Srivastava, V., Singh, A.K., Kishor, A., Pal, N.R.: A generalized family of constant orness ordered weighted averaging operators. IEEE Trans. Fuzzy Syst. (2024)
17. Torra, v.: Aggregation operators. In: On Fuzziness, pp. 691–695. Springer (2013)
18. Wang, Y.-M., Luo, Y., Hua, Z.: Aggregating preference rankings using OWA operator weights. Inf. Sci. **177**(16), 3356–3363 (2007)
19. Zeshui, X.: An overview of methods for determining OWA weights. Int. J. Intell. Syst. **20**(8), 843–865 (2005)
20. Xu, Z.S., Da, Q.L.: The uncertain OWA operator. Int. J. Intell. Syst. **17**(6), 569–575 (2002)
21. Yager, R.R.: On ordered weighted averaging aggregation operators in multicriteria decision making. IEEE Trans. Syst. Man Cybern. **18**(1), 183–190 (1988)
22. Yager, R.R.: Applications and extensions of OWA aggregations. Int. J. Man-Mach. Stud. **37**(1), 103–122 (1992)

23. Yager, R.R.: Families of OWA operators. Fuzzy Sets Syst. **59**(2), 125–148 (1993)
24. Yager, R.R.: Quantifier guided aggregation using OWA operators. Int. J. Intell. Syst. **11**(1), 49–73 (1996)
25. Yager, R.R.: Centered OWA operators. Soft. Comput. **11**(7), 631–639 (2007)
26. Yager, R.R.: Using stress functions to obtain OWA operators. IEEE Trans. Fuzzy Syst. **15**(6), 1122–1129 (2007)
27. Yager, R.R., Filev, D.P.: Parameterized and-like and or-like OWA operators. Int. J. General Syst. **22**(3), 297–316 (1994)
28. Yager, R.R., Filev, D.P.: Induced ordered weighted averaging operators. IEEE Trans. Syst. Man Cybern. Part B (Cybernetics) **29**(2), 141–150 (1999)

Open Access

Open Access This chapter is licensed under the terms of the Creative Commons Attribution-NonCommercial-NoDerivatives 4.0 International License (http:// creativecommons.org/licenses/by-nc-nd/4.0/), which permits any noncommercial use, sharing, distribution and reproduction in any medium or format, as long as you give appropriate credit to the original author(s) and the source, provide a link to the Creative Commons license and indicate if you modified the licensed material. You do not have permission under this license to share adapted material derived from this chapter or parts of it.

The images or other third party material in this chapter are included in the chapter's Creative Commons license, unless indicated otherwise in a credit line to the material. If material is not included in the chapter's Creative Commons license and your intended use is not permitted by statutory regulation or exceeds the permitted use, you will need to obtain permission directly from the copyright holder.

Tuning Echo State Networks with a Modified Elk Herd Optimizer for Improved Unemployment Rate Prediction

Muna Mohammed Al Mukhaini[1], Tamara Zivkovic[2], Miodrag Zivkovic[2], Snezana Anetic[2], Branislav Radomirovic[3], Csaba Varsandán[2], Luka Anicin[2], and Nebojsa Bacanin[2]([✉])

[1] Modern College of Business and Science (MCBS), Muscat, Oman
`muna.almukhaini@mcbs.edu.om`
[2] Singidunum University, Danijelova 32, 11010 Belgrade, Serbia
`{tzivkovic,mzivkovic,lanicin,nbacanin}@singidunum.ac.rs,`
`{snezana.anetic.24,csaba.varsandan.24}@singimail.rs`
[3] Institute of Artificial Intelligence, Belgrade, Serbia
`branislav.radomirovic@ivi.ac.rs`

Abstract. The economy serves as the foundation of any civilization, and a primary determinant that shapes it is the level of joblessness. The overall prosperity of a population is likewise impacted by this parameter, underscoring its critical significance. Conventional techniques for forecasting such economic patterns offer limited understanding and fall short of the capabilities enabled by artificial intelligence (AI). Consequently, the study outlined in this paper adopts an AI-based methodology to address this crucial challenge, which is governed by numerous intricate interdependencies. The task of forecasting unemployment levels is approached as a time-dependent data modeling problem, where recurrent neural networks (RNNs) often yield commendable outcomes. The investigation described in this work applies echo state networks (ESNs), a specific category within RNN architectures. Nevertheless, this strategy requires careful optimization of hyperparameters, leading to the integration of metaheuristic strategies as a practical remedy. To fine-tune the ESN configurations, an adapted variant of the elk herd optimization algorithm (ELK) is employed, tailored to meet the distinct demands of the problem. The suggested approach exhibits improved accuracy compared to both contemporary advanced techniques and the standard ELK, based on widely accepted evaluation metrics for temporal prediction.

Keywords: Echo state networks · Unemployment prediction · Elk herd optimizer · Metaheuristics optimization

1 Introduction

One of the fundamental obstacles to preserving a stable economic environment lies in the precise estimation of unemployment levels. This variable is widely

© The Author(s) 2026

J. C. Bansal et al. (Eds.): SCIS 2025, LNNS 1929, pp. 18–32, 2026.
https://doi.org/10.1007/978-3-032-22911-3_2

recognized as a crucial macroeconomic marker that reveals potential avenues for development, informs policymaking, and reflects the overall equilibrium of the financial system. Availability and reliable forecasting of this indicator is essential, as it allows prompt corrective measures and supports long-term resilience. Accurate prediction of unemployment dynamics also allows for early detection of both social and fiscal imbalances. Nonetheless, the emergence of influential elements such as gross domestic product (GDP) volatility, shifts in inflation and legislative adjustments introduces substantial intricacy to the task.

The capabilities of machine learning (ML) approaches have expanded significantly, especially in tackling time-dependent data problems. Within this scope, recurrent neural network (RNN) architectures have demonstrated noteworthy impact. Their recurrent structure facilitates the improved handling of sequential patterns [11]. However, conventional RNNs are now deemed obsolete, with more advanced and powerful architectures having taken their place. Among these, echo state networks (ESNs), gated recurrent units (GRUs), and long short-term memory (LSTM) networks [17] have emerged, each tailored to overcome inherent limitations of earlier RNN versions. Furthermore, when attention mechanisms are built in, the predictions become more reliable in sequential contexts, especially when paired with these enhanced RNN variants.

Despite their advantages, the efficiency of such architectures hinges on meticulous hyperparameter calibration, an inherently NP-hard challenge, signifying that deterministic algorithms are generally incapable of resolving it efficiently. This complexity arises from the necessity of extensive experimentation during the tuning process. As a solution, external optimization strategies based on heuristics are often adopted [13, 18]. Among these, metaheuristics inspired by collective intelligence, particularly swarm intelligence techniques, have delivered remarkable results in hyperparameter tuning scenarios. The investigation presented in this article employs the elk herd optimization algorithm (ELK) [2], a nature-inspired approach modeled after the collective movement of elk populations, known for its ability to navigate high-dimensional optimization landscapes. In addition, the study introduces a refined version of the baseline ELK algorithm, adapted to better suit the specifics of the unemployment prediction task via ESN tuning. The principal innovations introduced by this study are as follows:

- Development of a resilient AI-based model tailored for precise unemployment forecasting;
- Enhancement of the ELK metaheuristic to align with the unique demands of the addressed problem;
- Deployment of ESN architectures whose parameters are optimized using the refined ELK approach.

The rest of the paper continues with the following sections. Section 2 outlines the theoretical foundation of ESNs and related concepts utilized in this research. Section 3 elaborates on the modifications made to the original ELK algorithm. Section 4 details the setup of the experiments and configuration, while Sect. 5 presents and analyzes the results obtained. Finally, Sect. 6 concludes the manuscript and suggests avenues for future exploration.

2 Related Works and Background

Economic resilience and societal well-being are heavily reliant on the accurate estimation of joblessness trends. With the use of artificial intelligence (AI), major improvement has been achieved with conventional forecasting techniques and this segment focuses on reviewing related scientific efforts. Notable progress within the domain of recurrent neural networks (RNNs) has been marked by the introduction of long short-term memory (LSTM) units and gated recurrent units (GRUs). They were developed to achieve superior accuracy when modeling intricate temporal dependencies, which are inherent in time series prediction challenges. Effective modeling of unemployment rates has been achieved in the research by Yurtsever et al. [25], where the authors proposed a composite architecture combining LSTM and GRU units. Similarly, Yin et al. [24] explored the application of LSTMs for delineating lake boundaries, highlighting their adaptability. Furthermore, GRU networks have been utilized for estimating water quality in the study by Xu et al. [22], showcasing their versatility in handling various forecasting tasks.

In the realm of unemployment prediction, the integration of metaheuristic optimization techniques remains an underexplored strategy. For example, Sihombing et al. [9] applied particle swarm optimization (PSO) to refine an LSTM-based model for forecasting product sales. In another study, Syarif et al. [15] implemented symbiotic organism search to tune LSTM hyperparameters, this time targeting stock market data prediction, thereby underscoring the method's promise in enhancing model efficiency.

A fundamental computational hurdle in machine learning is the fine-tuning of hyperparameters, an endeavor widely acknowledged as a nondeterministic polynomial-time (NP-hard) problem. The inherent difficulty of this task is compounded by the implications of the no free lunch (NFL) theorem [21], which posits that no single optimization strategy can consistently outperform others across all types of problem, as model effectiveness is tightly coupled with dataset characteristics, evaluation metrics, and parameter configurations. To navigate these challenges, metaheuristic algorithms have gained notable traction in the research community. These approaches have produced impressive results in numerous domains, such as software development [18,19], medical diagnostics [26,27], and various applied sciences [3,4,10]. Nevertheless, their deployment in healthcare settings, especially those that involve complex and sensitive decision-making scenarios, remains comparatively rare [3,7,13].

The insights from the existing researches, it is evident that the utilization of echo state network (ESN) models in the context of unemployment rate forecasting is still not adequately addressed. This research aims to implement a specialized version of the elk herd optimization (EHO) algorithm to match the ESN parameters for a more precise modeling of the unemployment dynamics.

2.1   Echo State Network

Numerous variants of recurrent neural networks (RNNs) have been proposed in the literature [12], among which echo state networks (ESNs)[6] represent a distinct and noteworthy subclass. This particular architecture was developed to enhance the training efficiency typically associated with conventional RNNs. In ESNs, the temporal patterns within the input data are encoded through a dynamic structure known as the reservoir. Unlike traditional RNNs, only the output connections undergo the training process, which significantly reduces the overall computational burden. Furthermore, this architectural choice effectively mitigates issues such as vanishing and exploding gradients, thereby improving the stability of training. As a result, ESNs are particularly well suited for handling time-dependent prediction problems. The architecture of an ESN typically comprises an input layer, a large, sparsely connected reservoir acting as the hidden layer, and an output layer. The fundamental idea involves projecting the inputs into a high-dimensional space, supporting the identification of informative temporal features. At every time step, the input vector X_t is subjected to a linear transformation and combined with the previous reservoir state r_{t-1} to yield the current state r_t, as defined in Eq.(1).

$$r_t = \tanh(W_{in}X_t + W_{res}r_{t-1} + b_r), \tag{1}$$

In this formulation, W_{in} denotes the input weight matrix, while W_{res} represents the internal reservoir weight matrix. The term b_r corresponds to the bias applied to the reservoir state update. The nonlinear activation function used is the hyperbolic tangent tanh, which constrains the output values inside $[-1, 1]$.

The output vector Y_t is obtained using Eq. (2) as a linear combination of the current reservoir state r_t and the input vector X_t, weighted accordingly.

$$Y_t = W_{out}\left[r_t \; X_t\right], \tag{2}$$

Here, W_{out} denotes the output weight matrix, and the combined vector formed by concatenating the reservoir state and the input is represented as $\left[r_t \; X_t\right]$.

A fundamental distinction from standard RNNs lies in the fact that only the output weights W_{out} are subject to training. This adjustment is achieved by minimizing the error between the predicted outputs Y_t and the actual target values $\hat{Y}_t$, as defined in Eq. (3).

$$W_{out} = \operatorname{argmin}\sum_{t=1}^{T}|\hat{Y}_t - W_{out}\left[r_t \; X_t\right]|^2. \tag{3}$$

The echo state feature represents a fundamental characteristic of ESNs, ensuring that the impact of the initial reservoir state diminishes over time. This enables the network to move priority on the evolving patterns within the input sequence. The property is regulated by appropriately scaling the reservoir weight matrix. As the reservoir remains untrained and fixed, the computational cost is significantly reduced, making ESNs efficient for modeling sequential data.

3 Methods

3.1 The Baseline Elk Herd Optimizer

The ELK algorithm [2] is a recently introduced optimization technique based on swarm intelligence, modeled after the reproduction-related actions of elk populations. In the natural world, this process unfolds in two primary stages: the rutting period, during which the herd divides into multiple subgroups, each led by a dominant male, and the calving stage, where the leading bull and his harem produce offspring. In the original formulation of the algorithm, this biological behavior is governed by a single control variable, P, which sets the starting proportion of dominant bulls within the population. The algorithm begins by initializing a population of elks, consisting of bulls and their associated harems, which is mathematically represented by the matrix P in Eq. (4).

$$\mathbf{P} = \begin{bmatrix} x_1^1 & x_2^1 & \cdots & x_n^1 \\ x_1^2 & x_2^2 & \cdots & x_n^2 \\ \vdots & \vdots & \ddots & \vdots \\ x_1^N & x_2^N & \cdots & x_n^N \end{bmatrix}, \tag{4}$$

Within this matrix, N indicates the total number of individuals in the population and n corresponds to the dimensionality of each candidate solution. Each agent x^j is initialized following the rule specified in Eq. (5):

$$x_i^j = lb_i + (ub_i - lb_i) \times U(0,1) \tag{5}$$

where lb_i and ub_i represent the minimum and maximum values for the i-th dimension, and $U(0,1)$ denotes a uniformly distributed random variable over the interval $[0,1]$. Once initialization is complete, the individuals in the herd are sorted in ascending order based on their fitness values. Family units are then created according to the parameter B_r, resulting in a total of $B = |B_r \times N|$ families. The top-performing individuals are chosen as bulls, which then compete to establish their respective families. Harems are assigned to each bull using a roulette-wheel selection approach, where the probability p_j associated with bull x^j is proportional to its fitness score $f(x^j)$ relative to the combined fitness of all bulls, as outlined in Eq. (6).

$$p_j = \frac{f\left(\boldsymbol{x}^j\right)}{\sum_{k=1}^{B} f\left(\boldsymbol{x}^k\right)} \tag{6}$$

In the following calving stage, a new generation of individuals is produced, represented as $x_i^j(t+1)$. These descendants receive attributes from both their sire (the dominant bull x^{h_j}) and the corresponding harem member $x_i^j(t)$. If an inherited trait aligns with the index of the father, Eq. (7) is used for update.

$$x_i^j(t+1) = x_i^j(t) + \alpha \cdot \left(x_i^k(t) - x_i^j(t)\right) \tag{7}$$

Here, α is a randomly chosen value from the interval $[0, 1]$, which regulates the degree of influence exerted by a randomly selected individual $x^k(t)$ on the offspring. Larger α values introduce greater randomness, thereby promoting population diversity. In cases where the offspring is indexed the same way as the mother, the new candidate solution is formed by integrating elements from both the mother $x^j(t)$ and the matching bull $x^{h_j}(t)$, as outlined in Eq. (8):

$$x_i^j(t+1) = x_i^j(t) + \beta \left(x_i^{h_j}(t) - x_i^j(t) \right) + \gamma \left(x_i^r(t) - x_i^j(t) \right) \tag{8}$$

In this equation, $x_i^j(t+1)$ represents the i-th element of the j-th offspring at iteration $t+1$, h_j designates the bull leading the j-th harem, and r indicates the index of a randomly chosen bull. The parameters β and γ are random values drawn from the interval $[0, 2]$, and govern the relative influence of the father and an additional randomly selected bull, respectively. In alignment with real-world behavior, if a bull fails to effectively protect his harem, there is a chance that mating occurs with other bulls. In the final step, all bulls, harems, and generated offspring form a unified population, after which it is ranked ascending by their fitness scores, and the best-performing individuals carried forward for the subsequent iteration.

3.2   Proposed Modified ELK

The standard ELK algorithm is still broadly recognized as an effective optimization method across a wide range of modern problem domains. However, its capacity for exploration often diminishes during the later phases of the search process, potentially leading to early convergence. To address these flaws of the original ELK, an adjustable β-hill climbing strategy [1] is woven in. The corresponding mathematical expression for this improvement is presented in Eq.(9).

$$min\{f(x)|x \in X\} \tag{9}$$

Within the domain X, the collection of admissible solutions is defined as $x = x_1, x_2, \ldots, x_N$. Each individual solution x_i is restricted within the interval $[LB_i, UB_i]$, where LB_i and UB_i denote the minimum and maximum limits for the respective parameter. The objective function is represented by $f(x)$, and the population count is designated by N.

At the beginning, every x_i is initialized with a tentative candidate. Throughout the optimization process, two complementary mechanisms are employed to iteratively enhance these solutions. The parameter β expands the search space, facilitating broader global exploration, while η introduces random fluctuations that support local exploitation by refining solutions within their neighborhoods, as described in Eq. (10).

$$x_i' = x_i \pm rnd \times \eta \tag{10}$$

At this stage, a neighboring candidate x_i' is generated based on the current solution x_i, where rnd denotes a random scalar sampled with equal probability over the interval $[0, 1]$. The parameter η controls the magnitude of the perturbation; higher values lead to further exploratory steps. To facilitate this behavior, η is initially set to a large value to encourage wide-ranging exploration throughout the search space and is then progressively reduced to zero, by Eq. (11).

$$\eta_t = 1 - \frac{t^{\frac{1}{p}}}{T^{\frac{1}{p}}} \tag{11}$$

In this expression, t refers to the ongoing step count, T represents the total loop count, and the rate at which η decays is controlled by the parameter p. To ensure a gradual switch from global scouting to local fine-tuning, the value of p was set to 2 in this work. Once x_i' is generated, the parameter β is used to calculate the refined solution x_i'', incorporating the adjusted value of η as shown in Eq. (12).

$$x_i'' = \begin{cases} x_k, & rnd \leq \beta \\ x_i', & \text{otherwise} \end{cases} \tag{12}$$

In this representation, rnd refers to a uniformly distributed random number within the interval $[0, 1]$, and k corresponds to a randomly picked index from the range $[1, N]$. The current solution is stochastically replaced when rnd is less than or equal to β. The evolution of β over successive iterations follows a linear growth function, restricted between $\beta_{\min}$ and $\beta_{\max}$, which are initialized to .01 and .1, respectively, as defined in Eq. (13).

$$\beta_t = \beta_{\min} + \frac{t}{T} \times [\beta_{\max} - \beta_{\min}] \tag{13}$$

The outlined strategy is called adaptive βhill climbing (AHC), which has been integrated into the ELK metaheuristic framework. AHC is acknowledged for boosting the exploitation phase by adaptively adjusting the range across the search area. Therefore, it produces higher-quality solutions and improves precision in parameter fine-tuning. Additionally, the approach aids in avoiding local optima and enhances convergence speed during the later stages of the optimization process. Therefore, it reinforces both local refinement and global exploration, contributing to improved robustness, faster convergence, and greater accuracy in the final solutions. The upgraded version of the algorithm is termed hill climbing ELK (HCELK), and its operational steps are outlined in Algorithm 1.

4 Experimental Setup

This study relies on a freely available dataset at https://fred.stlouisfed.org/. The dataset comprises information on initial unemployment claims (ICSA), average hourly earnings for all employees (CES0500000003), and the Standard and Poor's 500 (S&P 500) index. Data entries are recorded weekly, covering the period from 2015 through December 2024. A noticeable surge appears amid the COVID-19 outbreak, serving as a solid illustration of how external shocks can impact the

---

Algorithm 1. Pseudocode of the proposed HCELK algorithm

Initialize the elk population P of size N using standard ELK initialization
Set the control parameters η and β for the adaptive hill climbing (AHC) component
while $t < T$ **do**
    Compute fitness values for all individuals in the population
    Execute the standard ELK search operations
    Apply the AHC enhancement procedure
    for each elk $x_i \in P$ **do**
        Create a neighboring candidate x_i' using Eq.(10)
        Sample a random number $rnd \sim U(0, 1)$
        if $rnd \leq \beta$ **then**
            Replace x_i with a randomly selected solution $x_k \in P$, ensuring $k \neq i$
        else
            Update x_i by assigning it the value of x_i'
        end if
        Recalculate η using Eq.(11)
        Adjust β based on Eq. (13)
    end for
end while
Output the best solution identified

---

labor market. To capture this effect, an auxiliary variable is introduced, assigned a value of 1 throughout the pandemic period. The data is broken down into 70% training, 10% validation, and 20% test sets, as illustrated in Fig. 1. Forecasts are performed one step ahead, using $lags = 8$.

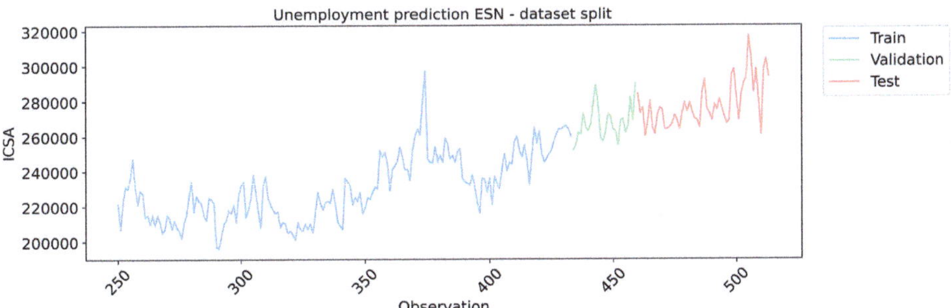

Fig. 1. Visualization of the dataset partitioning: (70%) training, (10%) validation, and (20%) testing.

The proposed HCELK algorithm was applied to tweak the parameters within the GRU network in the specified dataset. The set of competing methods included the basic ELK [2], variable neighborhood search (VNS) [14], firefly algorithm (FA) [23], sine cosine Harris Hawks Optimization (SCHO) [5], and COLSHADE [8]. Each of these algorithms was standalone coded in Python, utilizing the parameter settings recommended by their original authors. GRU optimization was carried out using populations consisting of eight agents, with six iterations allocated for improvement. To account for the stochastic nature of the algorithms and to establish a solid foundation for statistical evaluation, each method was independently executed thirty times. The search spaces and parameter ranges selected for the ESN tuning are presented in Table 1.

Table 1. Search space ranges for echo state network (ESN) hyperparameters used in the optimization process.

Limit	learning rate	dropout	number of epochs	ESN Layers	neurons
Lower	.0001	.05	50	1	50
Upper	.01	.2	100	3	100

The developed models are evaluated using a conventional suite of regression performance metrics [16]:

$$\text{RMSE} = \sqrt{\frac{1}{n}\sum_{i=1}^{n}\left(b_i - \hat{b}_i\right)^2} \tag{14}$$

$$\text{MAE} = \frac{1}{n}\sum_{i=1}^{n}\left|b_i - \hat{b}_i\right| \tag{15}$$

$$\text{MSE} = \frac{1}{n}\sum_{i=1}^{n}(b_i - \hat{b}_i)^2 \tag{16}$$

$$R^2 = 1 - \frac{\sum_{i=1}^{n}(b_i - \hat{b}_i)^2}{\sum_{i=1}^{n}(b_i - \bar{b})^2} \tag{17}$$

The index of agreement (IoA) [20] was also monitored during the optimization process to provide a broader perspective on model performance. The expression for this metric is given as follows:

$$\text{IoA} = 1 - \frac{\sum_{i=1}^{n}(b_i - \hat{b}_i)^2}{\sum_{i=1}^{n}(|b_i - \bar{b}| + |b_i - \hat{b}_i|)^2} \tag{18}$$

In the above equations, b_i and $\hat{b}_i$ denote the real and forecasted values for the i-th instance, respectively; $\bar{b}$ stands for the average of actual values and n represents the aggregate number of observations. The index of agreement (IoA) [20] was additionally monitored throughout the optimization process to offer a more holistic assessment of model performance. This inclusion serves a supplementary evaluation role, as the RMSE metric is used as the primary objective function for the optimization runs.

5   Experimental Outcomes

To begin with, the objective function results, measured using MSE, are shown in Table 2. In this evaluation, the proposed HCELK achieved the most effective values for both the best-case run (.007130) and the worst-case run (.007625). The original ELK algorithm achieved the lowest mean (.007338) and median (.007299) MSE values, with the proposed HCELK ranking close behind. In terms of consistency, baseline ELK also demonstrated least variable performance, evidenced by the tiniest standard deviation, and as a result, HCELK again performed as the next most stable method.

Table 2. Comparison of ESN simulation results based on the objective function (MSE) across different optimization algorithms.

Method	Best	Worst	Mean	Median	Std	Var
ESN-HCELK	**.007130**	**.007625**	.007367	.007497	.000213	4.53E-08
ESN-ELK	.007133	.007643	**.007338**	**.007299**	**.000195**	**3.82E-08**
ESN-VNS	.007537	.008582	.007969	.007907	.000407	1.66E-07
ESN-FA	.007699	.009035	.008392	.008758	.000560	3.14E-07
ESN-SCHO	.007159	.008633	.007917	.007660	.000578	3.34E-07
ESN-COLSHADE	.007134	.008210	.007710	.007808	.000468	2.19E-07

The graphical evaluation of the objective and indicator functions highlights the strong stability of the proposed HCELK, with only the baseline ELK demonstrating slightly more consistent results. These findings hold true for the fitness function (MSE) and for the indicator metric (R^2), as illustrated in Fig. 2.

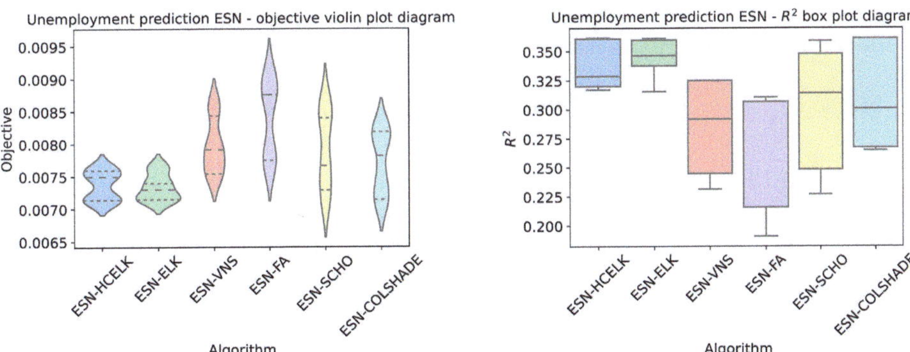

Fig. 2. Distribution comparison of objective function (MSE) and indicator function (R^2) values for all tested algorithms across multiple runs.

A more in-depth evaluation of the performance metrics, presented in Table 3, reveals that the introduced HCELK produced the optimal results in the categories R^2 (.361458), MSE (.007130) and RMSE (.084438). However, the baseline ELK yielded the highest performance for MAE and the index of agreement (IoA).

Table 3. Detailed evaluation metrics (MSE, RMSE, MAE, R^2, IoA) for the optimal ESN models obtained by each optimization algorithm.

Method	R^2	MAE	MSE	RMSE	IoA
ESN-HCELK	**.361458**	.065792	**.007130**	**.084438**	.740448
ESN-ELK	.361196	**.064521**	.007133	.084455	**.786564**
ESN-VNS	.325009	.068454	.007537	.086814	.781578
ESN-FA	.310506	.066181	.007699	.087742	.755253
ESN-SCHO	.358794	.066656	.007159	.084614	.760915
ESN-COLSHADE	.361044	.065251	.007134	.084465	.709935

The convergence curves of the objective and indicator functions for the best-performing execution of every algorithm are depicted in Fig. 3. The proposed HCELK attained the optimal value as early as the first iteration and maintained stability thereafter. The remaining algorithms also demonstrated satisfactory convergence behavior on this specific problem.

The discrepancy between the real and estimated values using the best-performing model of the proposed method for 1-step ahead forecasting is illustrated in Fig. 4.

The optimal configurations identified for each of the evaluated algorithms are listed in Table 4, where N/R denotes not relevant entry. This information is essential for reproducibility and provides a foundation for later researches.

Table 4. Selected hyperparameter configurations for the best-performing ESN models.

Method	Learning rate	Dropout	Training epochs	ESN Layers	Layer 1 Neurons	Layer 2 Neurons	Layer 3 Neurons
ESN-HCELK	.006978	.2	71	1	53	N/R	N/R
ESN-ELK	.008082	.164104	56	1	89	N/R	N/R
ESN-VNS	.006814	.131252	85	1	98	N/R	N/R
ESN-FA	.01	.05	74	1	66	N/R	N/R
ESN-SCHO	.01	.051602	100	2	50	61	N/R
ESN-COLSHADE	.01	.114147	53	2	64	82	N/R

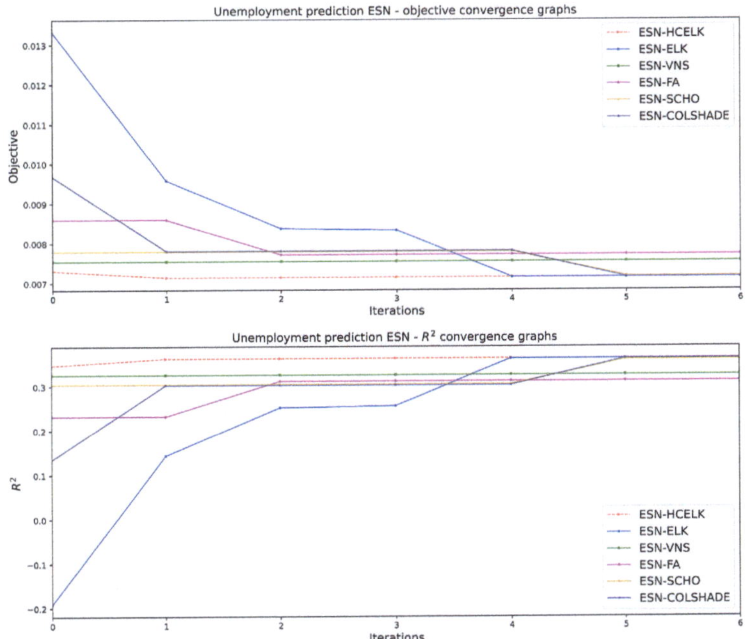

Fig. 3. Convergence curves of the objective function (MSE) and indicator function (R^2) for the best run of each optimization algorithm.

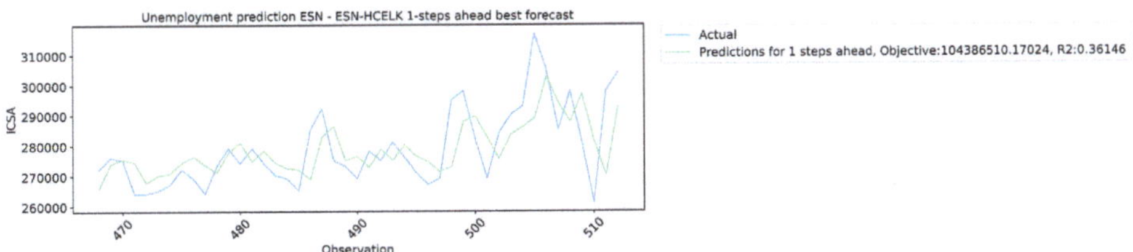

Fig. 4. One-step-ahead predictions of the best-performing ESN model optimized by the proposed HCELK algorithm, compared to actual values.

6   Conclusion

The unemployment rate holds a vital position in determining the overall stability and performance of the global economy. It not only shapes monetary shifts, but additionally influenced by numerous external variables that are not strictly tied to market dynamics. In this study, a publicly available dataset compiled from various sources was utilized, notable for including data from the COVID-19 pandemic, which is a compelling example of such non-market influences. To address the challenge of forecasting unemployment rates, a novel method is introduced that combines RNNs with metaheuristic optimization. The ESN is selected as the primary predictive model. Given that ESNs require hyperparameter optimization, the authors propose an enhanced version of the ELK algorithm, augmented with the AHC mechanism to improve exploratory performance. The proposed hybrid approach significantly outperformed all competing methods, demonstrating strong potential for this task. In addition to the new HCELK method, the original ELK and several cutting-edge optimization algorithms were evaluated, and HCELK achieving superior results in most cases. Upcoming research will concentrate on experimenting with alternative metaheuristics and exploring different predictive architectures.

Acknowledgments. Financial backing for this research came from the Science Fund of the Republic of Serbia, grant No. 7373, characterizing crises-caused air pollution alternations using an artificial intelligence-based framework (crAIRsis), and grant No. 7502, Intelligent Multi-Agent Control and Optimization applied to Green Buildings and Environmental Monitoring Drone Swarms (ECOSwarm).

References

1. Al-Betar, M.A., Aljarah, I., Awadallah, M.A., Faris, H., Mirjalili, S.: Adaptive β-hill climbing for optimization. Soft. Comput. **23**(24), 13489–13512 (2019)
2. Al-Betar, M.A., Awadallah, M.A., Braik, M.S., Makhadmeh, S., Doush, I.A.: Elk herd optimizer: a novel nature-inspired metaheuristic algorithm. Artif. Intell. Rev. **57**(3), 48 (2024)
3. Antonijevic, M., Jovanovic, L., Bacanin, N., Zivkovic, M., Kaljevic, J., Zivkovic, T.: Using Bert with modified metaheuristic optimized xgboost for phishing email identification. In: International Conference on Artificial Intelligence and Smart Energy, pp. 358–370. Springer (2024)
4. Bacanin, N., et al.: Crop yield forecasting based on echo state network tuned by crayfish optimization algorithm. In: 2024 IEEE International Conference on Contemporary Computing and Communications (InC4). vol. 1, pp. 1–6. IEEE (2024)
5. Bai, J., et al.: A sinh cosh optimizer. Knowl.-Based Syst. **282**, 111081 (2023)
6. Chatzis, S.P., Demiris, Y.: Echo state gaussian process. IEEE Trans. Neural Netw. **22**(9), 1435–1445 (2011). https://doi.org/10.1109/TNN.2011.2162109
7. Dobrojevic, M., et al.: Cyberbullying sexism harassment identification by metaheurustics-tuned extreme gradient boosting. Comput. Mater. Continua **80**(3) (2024)
8. Gurrola-Ramos, J., Hernandez-Aguirre, A., Dalmau-Cedeno, O.: Colshade for real-world single-objective constrained optimization problems. In: 2020 IEEE Congress on Evolutionary Computation (CEC), pp. 1–8 (2020)
9. He, Q.Q., Wu, C., Si, Y.W.: LSTM with particle swam optimization for sales forecasting. Electron. Commer. Res. Appl. **51**, 101118 (2022)
10. Lakicevic, B., et al.: Artificial neural networks with soft attention: natural language processing for phishing email detection optimized with modified metaheuristics. In: International Conference on Advanced Network Technologies and Intelligent Computing, pp. 421–438. Springer (2024)
11. Markovic, I., et al.: Flood prediction based on recurrent neural network time series classification boosted by modified metaheuristic optimization. In: International Conference on Advances in Data-driven Computing and Intelligent Systems, pp. 289–303. Springer (2023)

12. Medsker, L., Jain, L.C.: Recurrent neural networks: design and applications. CRC Press (1999)
13. Mladenovic, D., et al.: Sentiment classification for insider threat identification using metaheuristic optimized machine learning classifiers. Sci. Rep. **14**(1), 25731 (2024)
14. Mladenović, N., Hansen, P.: Variable neighborhood search. Comput. Oper. Res. **24**(11), 1097–1100 (1997)
15. Syarif, A., et al.: Hyperparameter optimization of long-short term memory using symbiotic organism search algorithm for stock prediction. Int. J. Innov. Res. Sci. Eng. Technol. **5**(2), 1–8 (2022)
16. Tatachar, A.V.: Comparative assessment of regression models based on model evaluation metrics. Int. J. Innov. Technol. Exploring Eng. **8**(9), 853–860 (2021)
17. Todorovic, M., Petrovic, A., Toskovic, A., Zivkovic, M., Jovanovic, L., Bacanin, N.: Multivariate bitcoin price prediction based on LSTM tuned by hybrid reptile search algorithm. In: 2023 16th International Conference on Advanced Technologies, Systems and Services in Telecommunications (TELSIKS), pp. 195–198. IEEE (2023)
18. Villoth, J.P., et al.: Two-tier deep and machine learning approach optimized by adaptive multi-population firefly algorithm for software defects prediction. Neurocomputing **630**, 129695 (2025)
19. Villoth, S.J., et al.: Optimizing error detection in generated code using metaheuristic optimized natural language processing. In: International Conference on Soft Computing and its Engineering Applications, pp. 239–253. Springer (2025)
20. Willmott, C.J., Robeson, S.M., Matsuura, K.: A refined index of model performance. Int. J. Climatol. **32**(13), 2088–2094 (2012)
21. Wolpert, D., Macready, W.: No free lunch theorems for optimization. IEEE Trans. Evol. Comput. **1**(1), 67–82 (1997)
22. Xu, J., Wang, K., Lin, C., Xiao, L., Huang, X., Zhang, Y.: Fm-GRU: a time series prediction method for water quality based on seq2seq framework. Water **13**(8), 1031 (2021)
23. Yang, X.S., He, X.: Firefly algorithm: recent advances and applications. Int. J. Swarm Intell. **1**(1), 36–50 (2013)
24. Yin, L., et al.: U-net-LSTM: time series-enhanced lake boundary prediction model. Land **12**(10), 1859 (2023)
25. Yurtsever, M.: Unemployment rate forecasting: LSTM-GRU hybrid approach. J. Labour Market Res. **57**(1), 18 (2023)
26. Zivkovic, M., et al.: Ocular disease diagnosis using CNNs optimized by modified variable neighborhood search algorithm. In: International Joint Conference on Advances in Computational Intelligence, pp. 99–112. Springer (2024)
27. Zivkovic, M., Bacanin, N., Zivkovic, T., Jovanovic, L., Kaljevic, J., Antonijevic, M.: Parkinson's detection from gait time series classification using LSTM tuned by modified RSA algorithm. In: International Conference on Communication and Computational Technologies, pp. 119–134. Springer (2023)

Open Access This chapter is licensed under the terms of the Creative Commons Attribution-NonCommercial-NoDerivatives 4.0 International License (http://creativecommons.org/licenses/by-nc-nd/4.0/), which permits any noncommercial use, sharing, distribution and reproduction in any medium or format, as long as you give appropriate credit to the original author(s) and the source, provide a link to the Creative Commons license and indicate if you modified the licensed material. You do not have permission under this license to share adapted material derived from this chapter or parts of it.

The images or other third party material in this chapter are included in the chapter's Creative Commons license, unless indicated otherwise in a credit line to the material. If material is not included in the chapter's Creative Commons license and your intended use is not permitted by statutory regulation or exceeds the permitted use, you will need to obtain permission directly from the copyright holder.

A Distance Metric Based Strategy for Detection of Interest Flooding Attack (IFA) in Information-Centric Networking (ICN)

Kumari Nidhi Lal[1(✉)], Krishna Malani[2], and Advait Gadekar[1,2]

[1] Department of Computer Science and Engineering, Visvesvaraya National Institute of Technology (VNIT), Nagpur, India
`nidhilal@cse.vnit.ac.in`, `advaitgadekar3@gmail.com`
[2] Shri Ramdeobaba College of Engineering and Management Nagpur, Nagpur, India
`krishnamalani77@gmail.com`

Abstract. Nowadays, a rapid demand of content access leads to requirement of shifting towards content-centric infrastructures from IP-based Internet framework. Due to its scalability and less latency of access data, it makes it more predominant to be used along with cloud computing, fog computing, and edge computing. The features like server hit reduction, load balancing, and low computation make Information-centric Networking (ICN) an attention seeker in the eyepoint of various service providers like AWS. For content access, ICN does not depend on tracking of IP addresses; instead, it can provide the content to users by utilizing the nearby Content Routers (CRs). This mechanism leads to various serious security threats like flooding attacks, denial of service attacks, and content poisoning attacks in the system, which lead to resource exhaustion and performance degradation due to malicious activity. Motivated by this, in this paper, an approach has been proposed for the detection of flooding attacks in ICN. For content access, the client initiates an interest packet in the network; when this packet is mistreated by malicious users, i.e., attackers, to launch a flooding attack, it is termed as Interest Flooding Attack (IFA). The proposed approach utilizes a distance metric-based scheme called Jensen-Shannon Divergence (JSD), which smoothly adapts the dynamicity of the IFA attack patterns in the network. In state-of-the-art, various strategies like Kullback-Leibler Divergence have been implemented and compared to the JSD approach. The resultant outcomes of plots illustrate that our anticipated approach performs preeminently than existing distance metric-based approaches in the detection and identification of IFA in ICN, in terms of lightweight implementation and variability of fluctuations in attack patterns.

Keywords: Information Centric Networking · Interest flooding attacks · AI-based Routing · Adaptive Routing

© The Author(s) 2026
J. C. Bansal et al. (Eds.): SCIS 2025, LNNS 1929, pp. 33–43, 2026.
https://doi.org/10.1007/978-3-032-22911-3_3

1   Introduction

Nowadays, the need of accessing data/content is more important than its access mechanism. The users on the Internet are now around 6 billion, as per the data published in 2025 [8]. In this term, the exhaustion of IP addresses somehow solved by IPV6, but still suffers from delayed access and high overhead on the server for each time content retrival. This problem is solved by the ICN architechture by concentrating only on the content items rather than their location of access. This flexibility is provided by introducing some cache capacity to intermediate routers in the path from client to server. In Information-Centric Networking, each content router maintains three core entriy tables: 1) Pending Interest Table (PIT) 2) Content Store (CS) 3) Forwarding Information Base (FIB) [12]. The PIT records incoming Interest requests received from client, while the CS temporarily caches content objects to enable faster retrieval. The FIB, on the other hand, stores the forwarding path for incoming Interests by mapping content names to appropriate outgoing interfaces. In ICN has two fundamental packet types for content request and delivery. First is Interest packet, it is issued by a client to request specific content based on its name and associated metadata. Once a matching Interest is satisfied, Data packet is generated by the content producer or an intermediate cache and returned to the client and carries the requested data payload. [11].

The functionality of the network will suffered if any malicious activity happened in the form flooding attack. In ICN, the flooding attack is performed by an attacker by generating the fake numerous Interest packet by impersonates the ID of a legitimate user in the network. In a fake Interest packet, the attacker appends bogus meta information of the content. This bogus information can be like a forged content name which is not actually exist in the cache of CR or the server. It results, choking the PIT of target CR in such a way that after some time, it will not be able to serve any incoming interest packet either forged or legitimate. In this ways, the Interest packet generated by client are not satisfied, and it leads to high Not or Negative Acknowledgment (NACK) packets in the network due to fake content that is not found in the CS of the CR and server. In addition, it degrades the cache hit and server hit reduction of the network [1].

Motivated by this, an IFA detection mechanism has been proposed for ICN that is based on the feature of the distance metric between the normal network profile and the attacked network profile. The summary of the contributions in this paper as follows:

- First, the IFA's impact on the ICN network has been evaluated and analyzed. Various performance metrics, for example, cache hit ratio, NACK packets, Interest drop rate by CR upon filling the PIT table, and latency, have been formulated and simulated in a network simulator. Comparison is made among the metrics' performance under IFA and normal scenarios.
- Afterwards, a JSD based IFA detection approach has been proposed in which the metric of mean and variance fluctuations in various features of the network traffic has been monitored and detect the propbablity of a suspected attack.

- To validate the suitability and accuracy, various existing approaches have been implemented and compared with the JSD method.

2   Related Work

In [17], the authors stated that in order to detect IFA, a scheme based on a threshold to count fake interest packets can be used. This type of solution comes under the category of statistical approaches. However, the solution works fine with a static count of metrics and will not support the dynamicity of the attack pattern under different scenarios. In the rapid shift towards an Artificial Intelligence-based approach, various machine learning and deep learning solutions is explored by authors in [16]. The aim of these approaches is to detect the fluctuation patterns and classify the network under attack and normal in terms of accuracy. The solution provided in attack scenarios gives higher accuracy but lead to the overfitting problem due to high variance in the dataset when it is exposed to changes in the attack pattern of network traffic [2].

In [13], a variant of IFA named as collusive IFA (CIFA) has been detected by assuming that network traffic follows the pattern of time series data. Authors involved both the IFA and CIFA in the network scenario, and detection is done by "word extraction for time series classification". However, the schemes can be a suitable solution for time series-related network traffic. Still, the research gap lies with the detection of rapid fluctuations in different attack scenarios of energy-efficient content distribution [4]. In [14], a "reconstruction forest-based detection method" (RFDM) has been proposed by the authors to accurately and effectively determine the traffic fluctuations. This strategy aims to reduce the false positives, which directly impact upon accuracy of classification [10]. This is done by the calculation of reconstruction error in the phase of reconstruction of data, and restricts the malicious spoofed Interest packet forwarding in the network. RFDM is evaluated in terms of speed, accuracy, and resistance towards IFA. However, state-of-the-art approaches are only baseline approaches like Gini-impurity that are not able to provide the actual resistance score of the RFDM approach towards IFA.

In the literature review, various research gaps have been identified. Motivated by this, a novel approach has been proposed, named JSD for IFA detection in ICN. The JSD approach is a statistical measure using a distance metric between the normal and attacked profiles of the network. The overall aim of the JSD method is to find the dissimilarity between the training and testing profile distribution under network traffic. After analyzing the features of simulated network traffic like Interest drop rate, NACK rate, and latency, JSD approaches the detection by finding the subtle anomalies that are susceptible to IFA. It can be termed that the proposed approach is robust with sudden changes in network traffic and maintains a balance between efficiency and accuracy. These outcomes make the proposed approach suitable for ICN security enhancement strategies.

3 Proposed Work

To deal with the severity of IFA in ICN, various survey papers have been published by researchers in the literature [15]. Motivated by the comprehensive reviews, the primary focus of this research work is to implement a distannce statistical strategy named JSD, which is used to quantify the similarity between the training and testing profile distributions [7]. The approach starts work with first generating the baseline profile known as normal network traffic patterns, and measures the distance based on a mathematical model with respect to incoming real-time traffic attributes. If any divergence in terms of fluctuations has been detected, the proposed method indicates it as a statistical anamoly detection and generates the signal alarm of susceptible IFA in the network [3]. The proposed strategy can be termed as a solution to various defense mechanisms from IFA in a traditional way that suffers from a high number of false positives triggered by large legitimate traffic in a short interval of timestamp [5]. In addition, Artificial intelligence-based detection methods only analyze the network traffic generated after a fixed time interval to collect a good quantity for training and testing. In this way, Machine and Deep learning approaches might ignore the temporal and spatial relationship between features of the network traffic dataset [6].

The proposed approach is made to overcome the limitation of existing solutions by inheriting its foundation, which is statistical and lightweight due to low computation overhead. These features make the JSD method suitable for real-time monitoring of traffic features directly on hardware by using Digital Twins. JSD method does not use a discrete count; rather, it considers the entire probability distribution of network traffic features. In this way, the proposed method can effectively determine the normal traffic from the attacked one during the flooding event intervals.

3.1 Interest Flooding Attack

In launching the IFA in the network, malicious users exploit the basic communication packets known as Interest packets by flooding them in the network. This flooding Interest packet contains requests for spoofed or fake content names that are non-existent in the CS of CRs and the server. It results in a massive amount of unsatisfied spoofed Interest packets flooding the network. The target of these spoofed Interest packets is to overflow the storage space of the PIT table and which leads to an increase of overhead in CR as well as a high drop rate of incoming legitimate interest packets [1]. The proposed approach is symmetric and able to detect real-time IFA because it is bound between 0 and 1. The normal traffic of the network is determined by implementing various metrics like PIT occupancy, Interest packet drop rate, Interest Request rate, Interest satisfaction rate, and NACKs. Afterwards, JSD metric is calculated based on a mathematical equation and compared with continuously coming real-time network traffic. If the computed metric diverges from the predefined threshold, then an IFA alarm signal as an alert will be generated by the proposed strategy. This alarm indicates the network might be suffering from potential IFA.

3.2   Distance Metrics for Anamoly Detection in IFA

To assess the potency of the intended work, various existing techniques are examined for making comparison. This section elaboartes various distance metric-based methods using mathematically formulation and applied in order to detect IFA in ICN.

1. Bhattacharyya Distance (BD): If the dataset points follow the normal distribution, Bhattacharyya Distance provides good accuracy in terms of divergence detection from normal to attacked profiles of network features. Mathematically, it can be computed using Eq. 1:

$$D_{BD}(A \parallel B) = -\ln\left(\sum_{i \in I} \sqrt{A(i)B(i)}\right) \tag{1}$$

Here, noraml profile is denoted by A and B indicates and attacked traffic distributions.

2. Mahalanobis Distance (MD): It quantifies the deviation of a point i from the mean m of a dataset by incorporating the covariance matrix Σ by capturing correlations among features. This property makes it suitable for multivariate anomaly detection scenarios where multiple traffic attributes, such as Interest request rates and NACK frequencies, are involved. It is formulated as Eq. 2.:

$$D_{MD}(i, m) = \sqrt{(i - m)^T \Sigma^{-1}(i - m)} \tag{2}$$

Although MD effectively extracts the correlation among the multiple feature but it dependence on accurate covariance estimation, which make it statistically extravagant, particularly in multivariate datasets.

3. Total Variation (TV): It measures the variability distance of two probability mass functions and can be expressed as Eq. 3:

$$\beta(A, B) = \frac{1}{2} \sum_{i \in I} |A(i) - B(i)| \tag{3}$$

It offers simple and intuitive indication of distributional divergence and varies between 0 and 1. However, despite its simplicity, it may fail to capture small fluctuations in probability distributions.

4. Kullback-Leibler (KL) Divergence: It quantifies the amount of information lost when the distribution Q is used as an approximation of P. It is represented by Eq. 4:

$$D_{KL}(A \parallel B) = \sum_{i \in I} A(i) \log \frac{A(i)}{B(i)} \tag{4}$$

Despite its high sensitivity to minor shifts between distributions, KL strategy is skewed and underperformed when $B(i) = 0$ and $A(i) > 0$, which restricts its relevance in sporadic traffic network conditions.

5. Jensen-Shannon (JSD): It is a symmetric and streamlined extension of KL metric which makes it well suited for attack detection in real time-span. It is formulated as Eq. 5:

$$D_{JS}(A \parallel B) = \frac{1}{2}D_{KL}(A \parallel \mu) + \frac{1}{2}D_{KL}(B \parallel \mu) \qquad (5)$$

where $mu = \frac{1}{2}(A + B)$.

3.3 IFA Detection in ICN Using JSD

For implementation, the probability distributions are needed in terms of A and B. Therefore, the simulated network traffic must be converted into a probabilistic numerical format. It leads to defining four key attributes as follows:

- **Total number of requests:** It is calculated by count of Interest packets which are issued by clients and received/forwarded by the CRs during each iteration within a fixed interval.
- **Attack Interests:** This metric quantifies the volume of malicious or spoofed Interest packets generated by an attacker
- **Dropped Packets:** It measures the count of unsatisfied Interest packets dropped by CRs in the PIT table due to overflow.
- **NACK Sent:** This indicates the number of Negative or No Acknowledgement (NACK) packets sent by the server in response to content not found in storage.

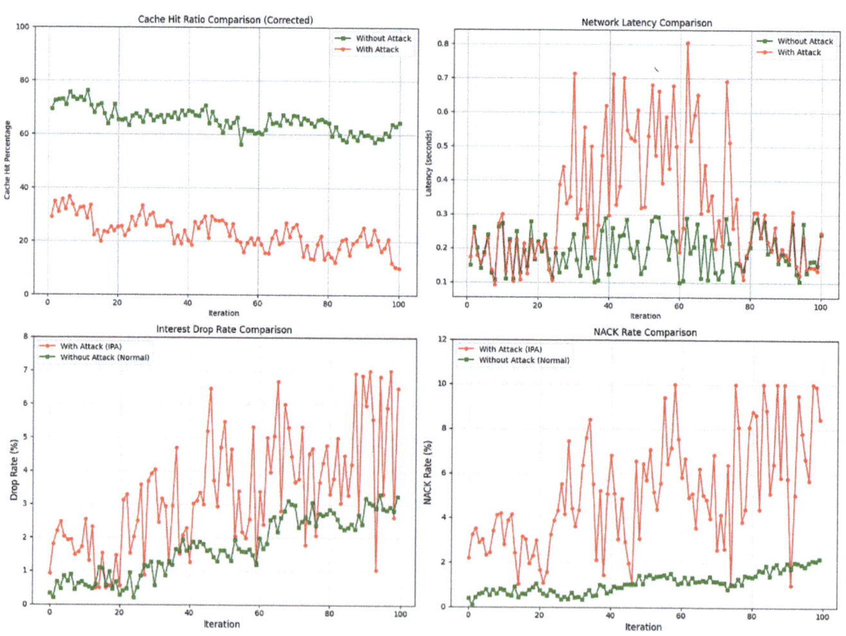

Fig. 1. Impact of IFA upon Network.

After formulating the above four attributes, in each time interval, the process of normalization will start. The computation of average values for all calculated and simulated four attributes will be followed by normalization to create the profiles of traning and testing probability distribution.

Let C = [Interest Requests count, spoofed Interest count, Dropped/loss Packets count, NACK packet count] be the vector of ICN simulation outcomes which are computed and normalized feature values. A_i is computed using Eq. 6 as:

$$A_i = \frac{C_i}{\sum_{j=1}^{4} C_j} \tag{6}$$

4   Performance Results

4.1   Simulation Setup

For simulation results, we have used own copyrighted simulator "SW-20011/2025" which is written in python. To generate the simulation environment, 10 content routers (cache capacity of 15 slots per router), 8 clients and 2 content servers have been taken. Content of 1000 request pattern have been generated by zipfs distribution $\alpha = 0.8$. The simulator also enables iterative experimentation under diverse conditions based on changing cache capacity, Interest generation rates and content popularity distributions [9].

4.2   Impact upon ICN Performance Metrices During IFA

In Fig. 1, network performance can be seen with resepect to some important parameters consideration. It can be seen that rapid drop of cache hit ratio is present when ICN network suffers from IFA attack. The value states that, ICN loss its feature of providing the content to user by utilizing nearby CRs. Instead, CRs of ICN are busy to balance the fake interest packet and leads to Denial of Service (DoS) attack. By using the numerical values present in the simulation, line plots shown in Fig. 1 are plotted to indicate the variation of various performance metric under the normal and attack scenario.

4.3   Detection of IFA in Network Using Distance Metrics Based Strategy

By using the simulator and setting network configurations, multiple iterations using different values of attack power and interest packets are simulated. After the simulation, the dataset is generated and for the feature engineering, it is divided into two set of profiles: 1) training 2)testing. For the feature enginerring, the score of feature importance have been evaluated. The important features are given to as input for the implementation of distance metrices scheme and plots are genereated indicated by Fig. 2a. Based on the score of feature importance, the implemented metrices are used to analyze the dependancy and relationship among them and calculates the probability of IFA in network. The classfication

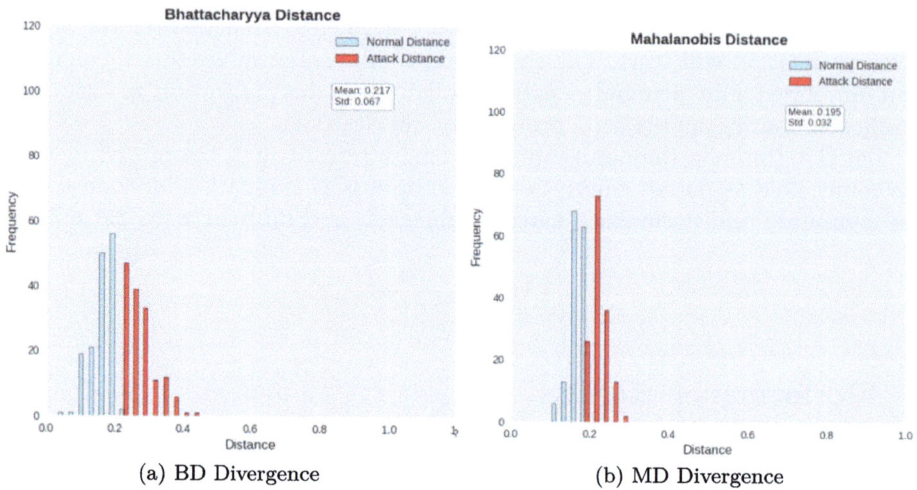

(a) BD Divergence   (b) MD Divergence

Fig. 2. Dissemination of BD and MD Distance metrics.

is done based upon the crossline of the threshold. It implies, the score of variation of features is calculated with respect to both the profiles in time interval t. If the computed score is exceeded the threshold, the strategy classified it as IFA else normal under the respective time interval instance. The Fig. 2b indicates that, Cumulative Distribution Function (CDF) of MD strategy is low in comaparision to BD scheme. In this regard, MD approach is not well suitable to detect IFA in ICN scenarios because lower value of CDF is not able to capture the non-linear and temporal features of the attack pattern.

We further evaluated the performance of the proposed work by applying multiple distance metrics to measure the divergence between training and testing

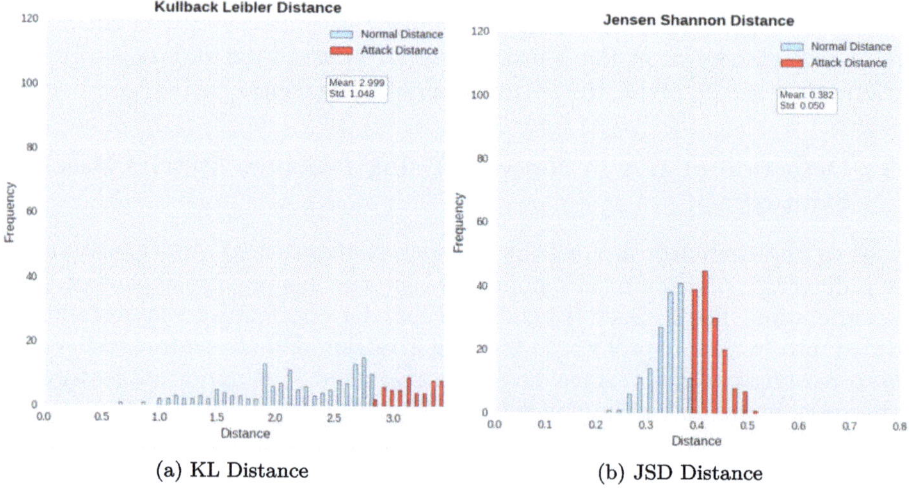

(a) KL Distance   (b) JSD Distance

Fig. 3. Dissemination of KL and JSD Distance metrics.

profiles across different time intervals. Figure 3a and Fig. 3b illustrate KL distance and JSD distance respectively for each interval, where red bars indicate intervals identified as attack periods and blue bars indicate to normal traffic conditions. Among the evaluated methods, the Jensen – Shannon – based approach gives the highest mean distance value of 0.3823 compared to the other metrics. It uses a probabilistic method to capture the feature extraction and computation of anomaly detection. The Jensen – Shannon strategy demonstrates enhanced sensitivity to anomaly patterns. As a result, the proposed detection framework achieves a lower rate of false positive as shown in Fig. 3.

5   Conclusion and Future Scope

The intended work introduces a statistical approach for detecting IFA in data centric infrastruture paradigm like ICN by utilizing distance-based computation to quantify sensetive deviations among probability distributions of training and testing noramal combined attack traffic profiles. The results validate that anticipated statistical approach constructively detects divergence of normal traffic in attack scenarios. In addition, it maintains a balance between detection and computation overhead due to its statistical nature and lightweight. From the plotted graphs and numerical results, it can be illustrated that the JSD approach outperforms in comparison to other approaches, like KL divergence, in the scenarios of dynamic IFA attack patterns over the ICN scenarios. In future, this work will be enahnced by utilizing machine learning based dynamic methods to detect IFA in real-time. Further, a mitigation approach is also planned to implement such that ICN will recover from the severe effects caused by IFA and the identification of the malicious user, which is the origin of the attack.

Acknowledgment. "This research is funded by the Anusandhan National Research Foundation (ANRF), Government of India, under Grant No. ANRF/ECRG/2024/ 001662/ENS. We express our sincere gratitude to ANRF for their financial assistance and unwavering support throughout the course of this project. Their commitment to fostering research and innovation has been instrumental in the successful completion of this work."

References

1. Ahmed, B., Kerrache, C.A., Lagraa, N., Mastorakis, S., Lakas, A., Tahari, A.E.K.: Interest flooding attacks in named data networking: survey of existing solutions, open issues, requirements, and future directions. ACM Comput. Surveys **55**(7) (2022): 1-37. https://doi.org/10.1145/3539730
2. Hussain, M.A., Mohsan, S.A.H., Muhammad, G., Abbasi, S.: A machine learning-based interest flooding attack detection system in vehicular named data networking. Electronics **12**(18) (2023): 3870. https://doi.org/10.3390/electronics12183870
3. Yabin, X., Gu, P., Xu, X.: Research on Detection Method of Interest Flooding Attack in Named Data Networking. Intell. Autom. Soft Comput. **30**(1) (2021). https://doi.org/10.32604/iasc.2021.018895

4. Uichin, L., Rimac, I., Hilt, V.: Greening the internet with content-centric networking. In: Proceedings of the 1st International Conference on Energy-efficient Computing and Networking, pp. 179-182. (2010). https://doi.org/10.1145/1791314.1791342

5. Alberto, C., Conti, M., Gasti, P., Tsudik, G.: Poseidon: mitigating interest flooding DDoS attacks in named data networking. In: 38th Annual IEEE Conference on Local Computer Networks, pp. 630-638. IEEE, (2013). https://doi.org/10.1109/LCN.2013.6761300

6. Zhang, X., Li, R., Hou, W.: Attention-based LSTM model for IFA detection in named data networking. Sec. Commun. Networks **2022**(1), 1812273 (2022)

7. Maria, E.D., Schindelin, J.E.: A new metric for probability distributions. IEEE Trans. Inf. Theory **49**(7), 1858–1860 (2003). https://doi.org/10.1109/TIT.2003.813506

8. Barney, W.: Geographies of the Internet. In The Encyclopedia of Human Geography, pp. 1–13. Cham: Springer Nature Switzerland, (2025). https://doi.org/10.1007/978-3-031-25900-5_322-1

9. Nidhi, L., Kumar, S., Kadian, G., Chaurasiya, V.K.: Caching methodologies in Content centric networking (CCN): A survey. Comput. Sci. Rev. **31**, 39–50 (2019). https://doi.org/10.1016/j.cosrev.2018.11.001

10. Nidhi, L., Kumar, S., Chaurasiya, V.K.: A network-coded caching-based multicasting scheme for information-centric networking (ICN). Iranian J. Sci. Technol. Trans. Electr. Eng. **43**(3), 427–438 (2019). https://doi.org/10.1007/s40998-018-0171-4

11. Nidhi, L.K., Kumar, A.: An efficient lookup search and forwarding mechanism for information-centric networking (ICN). Arabian J. Sci. Eng. **43**(12), 6849–6861 (2018). https://doi.org/10.1007/s13369-017-2903-6

12. Nidhi, L.K., Kumar, A.: A centrality-measures based caching scheme for content-centric networking (CCN). Multimedia Tools Appl. **77**(14), 17625–17642 (2018). https://doi.org/10.1007/s11042-017-5183-y

13. Wang, D., Li, W., Hou, R.: Blending interest flooding attacks detection in named data networking, In: 2024 IEEE International Conference on High Performance Computing and Communications (HPCC), Wuhan, China, 2024, pp. 1341-1346, https://doi.org/10.1109/HPCC64274.2024.00179.

14. Guanglin, X., Li, X., Hou, R.: A reconstruction forest-based interest flooding attack detection method in named data networking. In 2024 International Conference on Computing, Networking and Communications (ICNC), pp. 823-829. IEEE Comput. Soc. (2024). https://doi.org/10.1109/ICNC59896.2024.10556003

15. Ogunbunmi, S., Chen, Yu., Zhao, Q., Nagothu, D., Wei, S., Chen, G., Blasch, E.: Interest flooding attacks in named data networking and mitigations: recent advances and challenges. Future Internet **17**(8), 357 (2025)

16. Manivannan, K., Abinaya, S., Jasmine, T.A., Arathi, P., Baskar, D.: A hybrid strategy for addressing interest flooding attacks in named data networking. In 2025 3rd International Conference on Artificial Intelligence and Machine Learning Applications Theme: Healthcare and Internet of Things (AIMLA), pp. 1-6. IEEE, (2025). https://doi.org/10.1109/AIMLA63829.2025.11040569

17. Krishna, K.M., Tripathi, N.: Detecting interest flooding attacks in NDN: a probability-based event-driven approach. Comput. Secu. **148**, 104124 (2025). https://doi.org/10.1016/j.cose.2024.104124

Open Access This chapter is licensed under the terms of the Creative Commons Attribution-NonCommercial-NoDerivatives 4.0 International License (http://creativecommons.org/licenses/by-nc-nd/4.0/), which permits any noncommercial use, sharing, distribution and reproduction in any medium or format, as long as you give appropriate credit to the original author(s) and the source, provide a link to the Creative Commons license and indicate if you modified the licensed material. You do not have permission under this license to share adapted material derived from this chapter or parts of it.

The images or other third party material in this chapter are included in the chapter's Creative Commons license, unless indicated otherwise in a credit line to the material. If material is not included in the chapter's Creative Commons license and your intended use is not permitted by statutory regulation or exceeds the permitted use, you will need to obtain permission directly from the copyright holder.

A Novel Hybrid Ensemble Architecture for Stroke Risk Prediction Using Healthcare Data

Tushar Ghosh[✉], Jossy George, and S. Chanti

Christ (Deemed to Be University), Bangalore, India
`tusharghosh408@gmail.com`, `{frjossy,chanti.s}@christuniversity.in`

Abstract. Stroke is the reason for an alarming number of disabilities worldwide, further emphasising the critical need for early and accurate prediction of risks to inform clinical management. This paper presents a novel hybrid ensemble architecture that leverages the superiority of multiple machine learning models for stroke health risk prediction using health data. In this novel hybridisation, decision tree classifiers belonging to the Random Forest and XGBoost families are effectively combined with support vector machines and a shallow neural network within a Stacked ensemble strategy that uses a hard vote technique. To improve model generalizability and avoid overfitting, feature selection and dimensionality reduction methods like Recursive Feature Elimination (RFE) and Principal Component Analysis (PCA) have been included expertly without compromising performance. After extensive training and testing on a real-world health repository covering a broad range of demographic, lifestyle, and clinical features, the model obtained an outstanding F1-score of 0.9427 and an exemplary ROC-AUC value of 0.9872, much higher than the performance of the individual models. Statistical significance was assessed using the Friedman and Wilcoxon signed-rank test. The model is a strong candidate for incorporation into clinical decision support systems and is fully deployable and EHR-compatible.

Keywords: Stroke Prediction · Ensemble Learning · Machine Learning · Healthcare Analytics · Hybrid Models · Classification · Feature Selection · PCA · Stacking

1 Introduction

Stroke is one of the major contributors to long-term disability worldwide. It affects almost 15 million people per year and places huge demands on the healthcare system [1]. Early identification of individuals at high risk is crucial, as early classification will help undertake effective preventive measures, enhance clinical decision support, and overall lead to better patient outcomes.

Conventional stroke risk profile models, such as the Framingham Stroke Risk Profile, are mainly based on statistical approaches along with a core group of risk factors

© The Author(s) 2026

J. C. Bansal et al. (Eds.): SCIS 2025, LNNS 1929, pp. 44–55, 2026.
https://doi.org/10.1007/978-3-032-22911-3_4

[2]. While these techniques all have clinical utility, their limitation in handling complex dependencies between multiple health features, especially when faced with either imbalanced or heterogeneous data, calls for newer and more adaptable options.

Machine learning (ML) has become a powerful alternative for clinical prediction tasks, including stroke risk stratification [4]. Decision trees, neural networks, as well as support vector machines are some of the algorithms that have been employed [11] to uncover nonlinear relationships in large-scale health records. However, single-model methods often have a tendency to overfit, have limited generalizability, and show variability of performance across subpopulations within the target population [5]. On the other hand, ensemble techniques mitigate these obstacles by combining multiple models to boost robustness and general predictive power of the models [6]. More recently, hybrid model approaches with mixed type learners and feature selection methods have been able to increase the interpretability and performance of models significantly [7].

Here, we present a new hybrid ensemble approach designed specifically for stroke risk prediction. This approach introduces a combination of decision tree classifiers in the form of Random Forest and Extra Trees, a support vector machine, and a shallow neural network in a stacked ensemble setup. To reduce the risk of overfitting while improving the performance of the model, dimensionality reduction as well as feature selection methods were used. Both Recursive Feature Elimination (RFE) and Principal Component Analysis (PCA) are acknowledged as methods utilised here. Assessment is done through a real-world healthcare data set that includes demographic, lifestyle, and clinical data. The main contributions of this work are outlined below. Firstly, we propose a hybrid ensemble model combining both linear and nonlinear classifiers in order to provide efficient stroke prediction. With its consensus-based feature selection pipeline and large-scale experimentation and benchmarking process, this model shows promising improvements in F1-score, accuracy, and ROC-AUC, signifying its potential efficacy and prospective clinical value in stroke risk prediction.

This paper follows the following format: Section 2 reviews prior work on stroke prediction using ML techniques. Section 3 explains our suggested methodology. Section 4 describes the dataset and the preprocessing steps. Section 5 provides an account of the results and the discussion. Section 6 performs statistical significance analysis in order to confirm the robustness of the model suggested. Section 7 highlights the conclusion and possible avenues for future work.

2 Literature Review

The application of machine learning in stroke risk prediction is increasingly common, as various studies have explored numerous algorithms and ensemble methods in hopes of improving accuracy and consistency.

Govindarajan et al. [1] have done an extensive review of hybrid modelling frameworks in stroke care, with a focus on the benefits of combining statistical and machine learning approaches.

Jayasuriya et al. [2] performed a systematic review of stroke risk prediction models, showing us how effective methods that handle heterogeneous and imbalanced healthcare data are.

Chakraborty et al. [3] suggest a stacked machine learning model with feature selection and data preprocessing. The model exhibits better prediction performance than a single classifier.

Mochurad et al. [4] merged XGBoost with principal component analysis (PCA) and artificial intelligence methods, immensely increasing the interpretability and precision of stroke risk predictions.

Thakre et al. [5] presented a strong hybrid deep learning model that integrates convolutional and recurrent neural networks for stroke detection, achieving remarkable outcomes on imaging-based datasets.

Dudhe [6] created an advanced machine learning framework that uses feature engineering and ensemble techniques to increase predictive accuracy.

Several other studies explored ensemble architectures that have combined diverse base learners.

Zhang et al. [7] developed a hybrid ensemble deep learning model for ischemic stroke prediction, demonstrating significant performance improvements.

Hashim et al. [8] developed a stacked ensemble for brain stroke prediction and showed its robustness across multi-source clinical data. In recent research, combinations of deep learning models with metaheuristic techniques have also been used for stroke diagnosis based on EEG and other physiological data.

Several research works such as Peng [9] and Liu [10] proposed a combination of classification-based and hybrid sampling method to deal with the class imbalance in Stroke datasets, which significantly improved reliability of prediction.

Along with these group strategies, Kumari et al. [11] compared some of the classical classifiers like Logistic Regression, Naive Bayes, SVM, and XGBoost in stroke prediction. Their paper emphasises the significance of hyperparameter tuning, reporting a highest accuracy of 92.87% with a tuned Naive Bayes model.

Potential works have referenced hybrid and ensemble approaches which have shown to be great at predicting strokes; however, there are still significant challenges in terms of trade-offs between model explainability, model generalizability across diverse populations, and learning from the class imbalanced distribution present in real healthcare data. Moreover, the vast majority of recent works rely on deep and (or) over-parameterised architectures notorious for loss of transparency and inappropriate deployability in a clinical environment. To alleviate this deficiency, our proposal consists in the development of a hybrid ensemble paradigm based on the combination of tree-based classifiers, support vector machines and shallow neural networks merged within the stacking framework guided by feature selection. Our paradigm enforces robustness, readability and clinical utility, while also maintaining high predictive accuracy on healthcare imbalanced data.

3 Methodology

This research utilises a systematic machine learning stroke risk prediction pipeline, with cutting-edge preprocessing data, feature selection, and a hybrid ensemble modelling style. The process follows best practices defined in recent publications [3, 4, 10], and confirmed using cross-validation and statistical techniques to provide robustness and generalizability. As Illustrated in Fig. 1, the proposed stroke prediction pipeline consists of preprocessing, feature selection, and ensemble modelling.

3.1   Data Description

This analysis uses a publicly available stroke data set, such as anonymised patient data with demographic, lifestyle and clinical details. Age, sex, hypertension, history of heart disease smoking status are some specific mentions (mean glucose is mentioned and not included in this list though it's not explicit if mean or N was used). The dependent variable is binary and it concerns the occurrence of stroke.

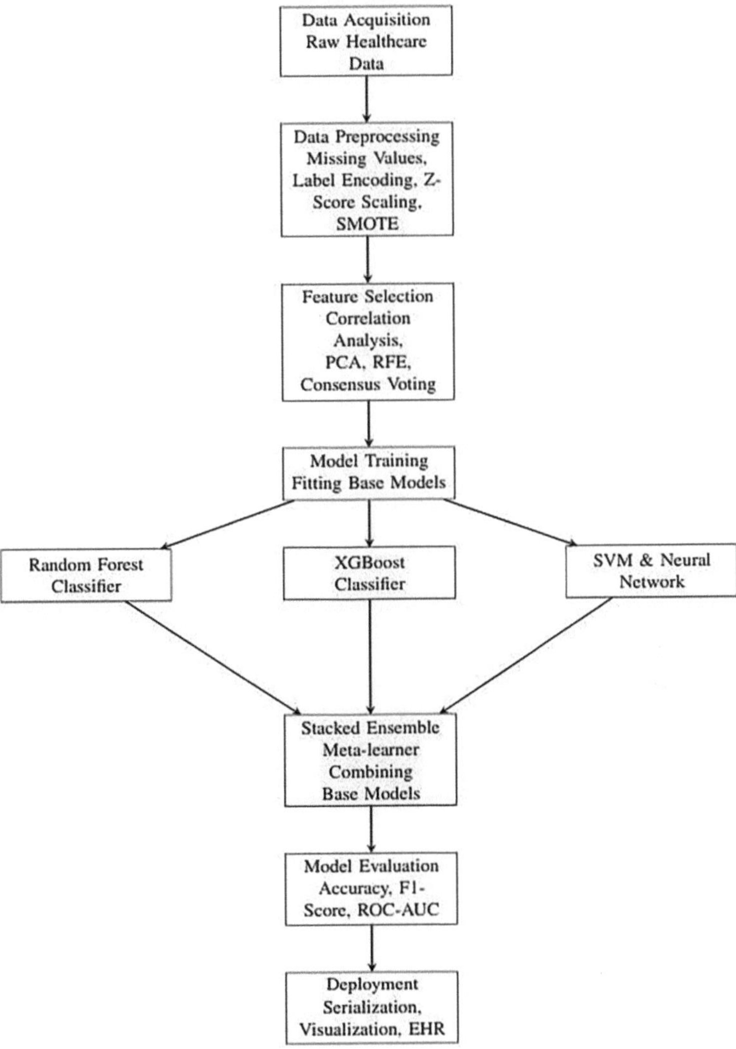

Fig. 1.   Workflow of the Proposed Hybrid Ensemble Stroke Prediction Pipeline

The exploratory data analysis (EDA) showed a stark class imbalance, where cases of stroke accounted for only 4.9% of the entire sample, as shown in Fig. 2. The imbalance complicated the model's learning process; hence, resampling methods were used in an effort to reduce bias. The distribution by age and stroke status (see Fig. 3) reinforces the point that stroke events are largely found in the older age groups, consistent with the prevailing clinical literature.

3.2 Data Preprocessing

A complete preprocessing pipeline was employed. to pre-process the data for machine learning operations:

- Missing Values: The Body Mass Index (BMI) column contained missing values, which were imputed using the mean.
- Encoding: Categorical variables such as gender, marital status, employment category, residence category, and smoking category. Were label-encoded to prepare them for model input.
- Scaling: Unbounded variables like age, mean glucose level, and BMI were z-score normalised to permit equal scaling.
- Imbalance Management: The Synthetic Minority Over-sampling Technique (SMOTE) was used to artificially produce samples from the minority (stroke) class, thus increasing the balance of the dataset and class imbalance reduction.

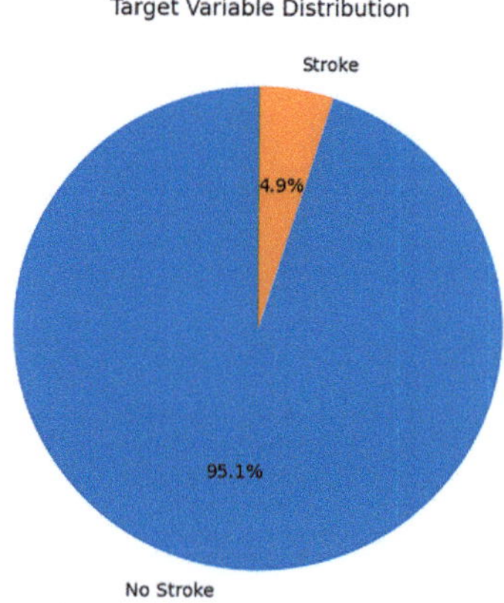

Fig. 2. Class distribution in the dataset showing stroke imbalance.

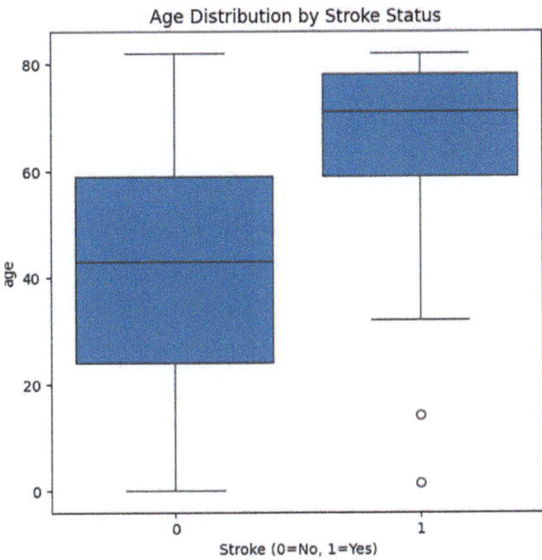

Fig. 3. Boxplot of age distribution grouped by stroke status.

3.3 Feature Selection and Engineering

To improve model generalisation and reduce redundancy, both domain-specific and statistical approaches to feature selection are used. A heatmap of correlation was used to determine multicollinearity among attributes, and strongly correlated pairs were either excluded or combined.

We also implemented Principal Component Analysis (PCA) to simplify feature dimensionality by retaining factors that explained 95% of total variance or more. Concurrently, Recursive Feature Elimination (RFE) by utilising Random Forests as the estimator was conducted to rank feature importance and incrementally remove low-contributing predictors. The final feature set was derived from the intersection of features selected by PCA and RFE.

In addition, domain knowledge was used to develop interaction-based attributes like a binary flag for patients with hypertension and diabetes. These engineered features were validated using cross-validation to ensure they improved predictive performance. They contributed positively to predictive performance. The frequency of feature selection across multiple statistical methods is illustrated in Fig. 4.

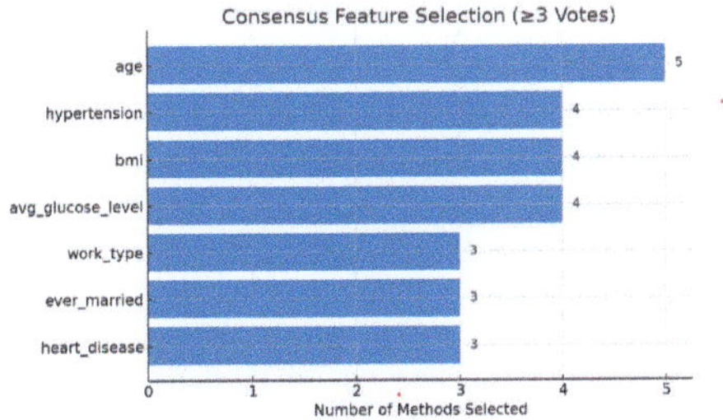

Fig. 4. Feature Selection Frequency Across Statistical Methods

3.4 Model Selection and Training

A few supervised learning algorithms were employed to determine whether they are applicable for predicting a stroke. The models assessed are Logistic Regression, Random Forest, Support Vector Machine (SVM), and eXtreme Gradient Boosting (XGBoost) for their effectiveness in health prediction tasks. We partitioned the dataset to two sets; training set and testing set with an 80:20 ratio. Training was performed on the predefined training set, while hyperparameter tuning was done by grid search cross-validation for performance tuning.

A stacked ensemble design was used where the baseline learners were SVM, Random Forest, XGBoost, and A single-hidden-layer neural network. (64 neurons). A Logistic Regression classifier was employed as the model and trained on the outputs of the base learners by using 5-fold cross-validation to avoid data leakage. Considering the disparity in counts between the stroke and no-stroke outcomes, we utilized the Synthetic Minority Over-sampling Technique (SMOTE) to train sensitive/specific models intended to be oriented toward sensitivity without harming specificity. The following model hyperparameters: the number of trees in Random Forest, learning rate in XGBoost and C-penalty parameter in SVM, were tuned by GridSearchCV with 5-fold cross-validation.

3.5 Evaluation Metrics

The performance of each classification model was tested. Rigorous tested against a fixed set of standard parameters appropriate for imbalanced binary results. These were accuracy, precision, Recall, F1-scores and Area Under the tongue Receiver Operating Characteristic curve. Characteristic Curve (ROC-AUC). You actually want Precision and Recall was particularly developed to account for the different cost of false positive or false negative in clinical practice. The classification results were illustrated with confusion matrix, and the performances of the models were measured in a ROC curve across different threshold value.

3.6 Model Deployment and Visualisation

To facilitate interpretability and usability, the last model was serialised using joblib for easy future inference. Visualisation software was used to present the findings effectively in an available presentation, i.e., feature importance plots, ROC curves, and prediction histograms. Interactive features were developed with Jupyter Notebook widgets to enable dynamic model behaviour exploration. These visualisations helped us find key risk indicators and showed insights into model decision pathways, hence enabling clinical interpretability and possible incorporation into decision support systems. The deployment environment was set up using Python 3.10, utilising scikit-learn and XGBoost libraries. The trained model is compatible with web-based inference APIs, thereby enabling potential Incorporation with Electronic Health Record (EHR) systems.

4 Experimental Setup

Experiments were run on a Windows 11 environment, along with an Intel Core i7 processor and 16 GB of RAM. The code was run in Python 3.10 with Jupyter Notebook. Core packages used were scikit-learn for model training and testing, XGBoost for gradient boosting, and imbalanced-learn for oversampling using the Synthetic Minority Oversampling Technique (SMOTE). Visualisation was done using matplotlib and seaborn. We divide the dataset into two categories, the training and testing sets, with a division of an 80:20 ratio. This ratio maintains the ratio of both stroke and non-stroke cases in the two sets. SMOTE was applied only to training data for class balancing without contaminating the test set. A five-fold cross-validation strategy approach was employed in model training to ensure stability and reduce variance.

Hyperparameters for every model were tuned by Grid-SearchCV. The following most important parameters were tuned:

- Random Forest: Num estimators, max depth, and minimum samples per leaf.
- XGBoost: Learning rate, number of estimators, and maximum depth.
- Support Vector Machine (SVM): Kernel type and regularisation parameter (C). The proposed architecture also pursued a stacked ensemble, made up of Random Forest, XGBoost, and SVM as models, using a Logistic Regression classifier as the model. Base model results were computed with out-of-fold predictions to prevent information leak during stacking.

The classification performance assessment on the initial test set for all models used a suite of criteria such as classification accuracy, precision, recall, F1-score, and Area Under the Receiver Operating Characteristic Curve (ROC-AUC). Random seed values were determined in a manner that ensured consistency and reproducibility across numerous iterations.

5 Results and Discussion

The last assessment entailed a comparison of ten models, comprising base classifiers, their corresponding tuned models, and six hybrid ensembles. It emerged that the top-performing model was the Hard Voting Ensemble that combines Random Forest and Extra Trees. Its evaluation metrics are reported in Table 1 for selected high-performing models for comparison.

Table 1. Test Set Performance of the Top Four Models

Model	Accuracy	Precision	Recall	F1-Score	ROC-AUC
Extra Trees (Original)	0.9404	0.9229	0.9609	0.9415	0.9862
Random Forest (Tuned)	0.9368	0.9199	0.9568	0.9380	0.9851
Stacked Ensemble	0.9419	0.9360	0.9486	0.9423	0.9872
Hard Voting (Proposed)	0.9424	0.9379	0.9475	0.9427	0.9872

The confusion matrix of the proposed Hard Voting model is shown in Fig. 5. The model correctly predicted 921 true positives and 912 true negatives, only 61 false positives and 51 false negatives. The low rate of false negatives is especially noteworthy—Early identification is critical in stroke prediction. The Precision-Recall curve shown in Fig. 6 is the model's capacity to maintain high accuracy over a spectrum of recall values. With an average precision (AP) of 0.99, the Hard Voting ensemble performs robustly under class Imbalance. Figure 7 is a scatter plot of ROC-AUC vs. F1-score over ensemble models. The Hard Voting model is positioned in the top-right quadrant, reflecting high discrimination and classification balance.

The suggested Hard Voting model outperformed the best base model (Extra Trees) in F1-score (0.9427 vs. 0.9415), accuracy (0.9424 vs. 0.9404), and ROC-AUC (0.9872 vs. 0.9862), though the relative gain is small (0.12%). Statistical testing confirmed the improvements observed, thus validating the stability of the suggested ensemble model. This emphasises the power of the group to control variance and optimise classification boundaries by combining different model strengths.

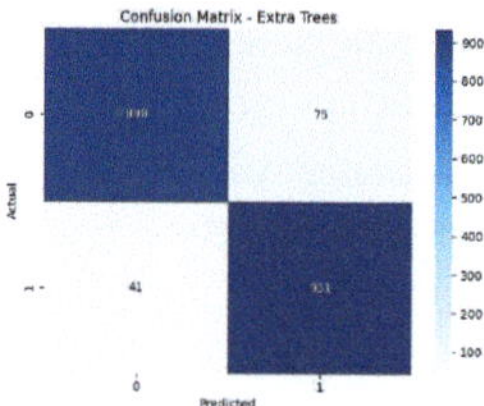

Fig. 5. Confusion Matrix of the Hard Voting Ensemble Model

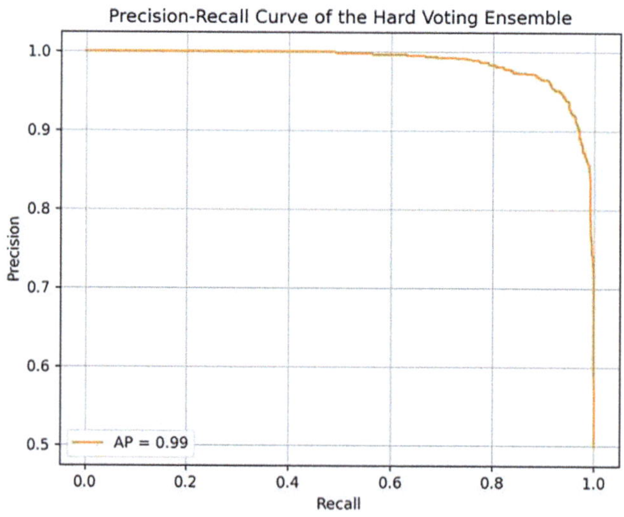

Fig. 6. Precision-Recall Curve (AP = 0.99) of the Hard Voting Ensemble

The hybrid ensemble proposed makes notable improvements in numerical results, but the improvements relative to some baseline models are marginal. This observation is not surprising given that many algorithms used such as the Random Forest and XGBoost are already performing particularly well. Thus, the key advantage is that not only are more accurate predictions made but also such models are robust, interpretable and have clinical usability along with a succinct feature selection mechanism. Moreover, the model's ability to maintain high performance under imbalanced data justifies its practical stability, which is just as important as striking numerical improvements in health prediction tasks.

6 Statistical Significance Analysis

This is to confirm whether the improvement observed from the proposed hard voting ensemble demonstrated statistical significance. Statistical hypothesis testing was employed in all the tests. Models employ their F1-score distributions. We used the Shapiro-Wilk test to determine normality at the beginning. The findings showed that F1-score distributions did not stray far from normality; therefore, they were worthy of the application of parametric and non-parametric tests. A Friedman test, an appropriate comparison with more than two comparable classifiers, was performed, and generated a statistically significant outcome ($p < 0.05$). This implies that not all aspects were favourable. For easier comparison of the differences, comparisons were made with the Wilcoxon signed-rank test.

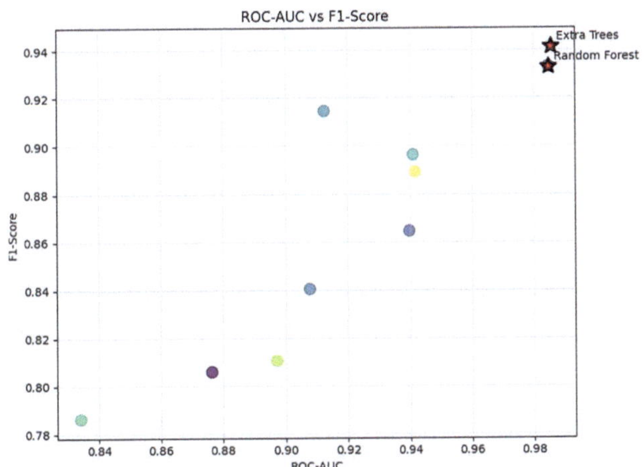

Fig. 7. Scatter Plot of ROC-AUC vs F1-Score for Ensemble Models

The findings showed that the Hard Voting ensemble greatly exceeded the performance of the foundational models, including Extra Trees and Random Forest analysis, which yielded a p-value of 0.016. These results offer statistical proof of the enhanced performance in question earlier, for both F1-score and ROC-AUC.

Table 2 presents the descriptive statistics of F1-scores across all models and forms the basis for subsequent statistical significance testing, while Fig. 8 illustrates a bar chart comparing their performance.

Table 2. Descriptive Statistics of F1-Score Distribution Across All Models

Metric	Score
Mean F1-Score	0.8680
Standard Deviation	0.0557
Minimum	0.7862
Maximum	0.9415

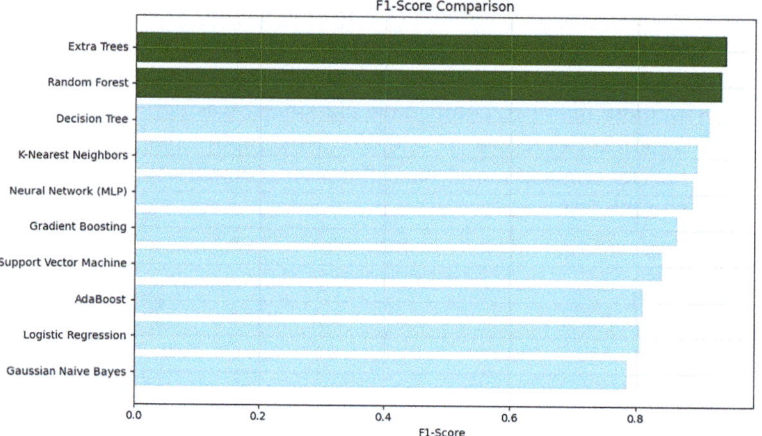

Fig. 8. Score Comparison of All Evaluated Models

7 Conclusion

In this paper, we introduced a hybrid ensemble method of tree-based classifiers and support vector machines for stroke risk prediction. The concluding Hard Voting ensemble, which combined Random Forest and Extra Trees, performed well with an F1-score of 0.9427 and ROC-AUC of 0.9872 on the test set, with a low false negative rate—a critical clinical measure. Statistical verification with the Wilcoxon signed-rank test validated the model's superiority over baselines. It also comes with limitations, although the results are promising. The dataset only accounts for a certain population and does not comprehensively depict the variability at the level of the region or the level of the hospital. Moreover, the absence of longitudinal or imaging data makes the idea of superior predictive capability suffer. Future works include the validation of the model by the application of multi-centre datasets and the integration of temporal features, so clinical generalizability can be improved. The findings support that ensemble approaches can improve the reliability of predictions in imbalanced healthcare datasets. The future can investigate the integration of temporal health records, the utilisation of deep learning-based ensembles, and improving interpretability using explainable AI methods for clinical uptake.

References

1. Govindarajan, N.S., Narayan, R., Vasan, S.: Hybrid modelling for stroke care: review and suggestions of new hybrid frameworks. Clin. Integrat. Care (2021)
2. Jayasuriya, S., et al.: Stroke risk prediction models: a systematic review and meta-analysis. J. Neurol. Sci. (2024)
3. Chakraborty, P., Bandyopadhyay, A., Sahu, P.P.: Predicting stroke occurrences: a stacked machine learning approach with feature selection and data preprocessing. BMC Bioinform. **25**, 329 (2024)
4. Mochurad, L., Babii, V., Boliubash, Y., Mochurad, Y.: Improving stroke risk prediction by integrating XGBoost, optimised principal component analysis, and explainable artificial intelligence. BMC Med. Inform. Decis. Mak. **25**, 63 (2025)
5. Thakre, G., Raut, R., Puri, C., Verma, P.: A hybrid deep learning approach for improved detection and prediction of brain stroke. Appl. Sci. **15**(9), 4639 (2025)
6. Dudhe, P.: An advanced hybrid machine learning framework for stroke risk prediction. Int. J. Innovat. Res. Technol. **12**(1), 5093–5098 (2025)
7. Zhang, X., Lee, K., Das, A., Patel, R.: Hybrid ensemble deep learning model for advancing ischemic brain stroke. Sci. Rep. **10**(7), 160 (2024)
8. Ukm, U., Hashim, T., Kaur, S.: Brain stroke prediction using stacked ensemble model. In: Proceedings of the International Conference on Stroke ML, Kuala Lumpur, Malaysia (2024)
9. Chen, Y., Li, F., Wang, J.: Hybrid deep learning and metaheuristic model-based stroke diagnosis. J. Biomed. Sign. Process. Control (2023)
10. Peng, G., Liu, J.: Enhanced stroke prediction through integrated classification and hybrid sampling techniques. In: Proceedings of the 3rd International Conference on Machine Learning, Cloud Computing, and Intelligent Mining, pp. 858–867. Guangxi, China (2025)
11. Kumari, P., George, J., Nair, A.M., Paul Alapatt, B., Baby, R., Jose, J.: Predicting and analyzing early onset of stroke using advanced machine learning classification technique. In: Proceedings of the 10th International Conference on Electrical Energy Systems (ICEES) (2024)

Open Access This chapter is licensed under the terms of the Creative Commons Attribution-NonCommercial-NoDerivatives 4.0 International License (http://creativecommons.org/licenses/by-nc-nd/4.0/), which permits any noncommercial use, sharing, distribution and reproduction in any medium or format, as long as you give appropriate credit to the original author(s) and the source, provide a link to the Creative Commons license and indicate if you modified the licensed material. You do not have permission under this license to share adapted material derived from this chapter or parts of it.

The images or other third party material in this chapter are included in the chapter's Creative Commons license, unless indicated otherwise in a credit line to the material. If material is not included in the chapter's Creative Commons license and your intended use is not permitted by statutory regulation or exceeds the permitted use, you will need to obtain permission directly from the copyright holder.

Scientometric Insights into Gesture Recognition Research Using Convolutional Neural Networks and Edge AI

Archana Reddy Rondla[1], Pasha Syed Nawaz[1], and Sai Anjani Manda[2]($\boxtimes$)

[1] Department of CSE, SR University, Warangal, Telangana 506002, India
`{r.archanareddy, sd.nawazpasha}@sru.edu.in`
[2] Center for Informetrics and Statistics, SR University, Warangal, Telangana 506002, India
`anjanitechno26@gmail.com`

Abstract. Gesture recognition has become the technology of improvement of human-computer interaction (HCI) using natural and touch-free communication interfaces. The paper provides a scientometric study of the world research on gesture recognition using convolutional neural networks (CNNs) and edge artificial intelligence (AI) between 2020 and 2025. The Scopus was used to collect data, and visualize collaboration networks, keyword patterns, and research evolution were done with VOS viewer and Biblioshiny. The research revealed 477 documents from 161 sources, whose annual growth rate stands at 8.45, which signifies the growing academic and industrial concern. The most prolific countries are China and India, and the main publishers are the IEEE Access, Sensors and IEEE Sensors Journal. CNNs, lightweight architectures, and multimodal fusion are emphasized as the main themes of co-occurrence analysis, but privacy-conscious AI systems based on edges and energy-efficient AI systems are becoming future research directions. The results show that there is a globally dispersed network of high author diversity and considerable interdisciplinary integration between computer vision, embedded systems, and artificial intelligence. This scientometric study offers information about intellectual organization, technological concentration, and research dissemination distribution of gesture recognition all over the world, and how CNN-driven and edge-enabled architectures are changing the future of intelligent, adaptable, and privacy conscious HCI applications.

Keywords: Gesture recognition · Convolutional neural networks · Edge AI · Human–computer interaction · Scientometric analysis

1 Introduction

Self-gesture recognition systems have emerged as an indispensable component in the advancement of human-computer interaction (HCI), offering intuitive, contactless, and natural modes for users to communicate with digital platforms and devices. Over the last decade, rapid developments in deep learning and artificial intelligence, particularly in convolutional neural networks (CNNs), have enabled significant improvements in the

© The Author(s) 2026
J. C. Bansal et al. (Eds.): SCIS 2025, LNNS 1929, pp. 56–68, 2026.
https://doi.org/10.1007/978-3-032-22911-3_5

accuracy and robustness of gesture recognition, allowing machines to interpret human movements with increasing nuance and efficiency, even in resource-constrained environments typical of embedded and edge devices Shyamala et al. [31]. Some converging trends, such as the ubiquity of low-power, high-capability embedded systems, the proliferation of a variety of sensory modalities, including RGB cameras, depth sensors, radar, and IMUs, and the push towards real-time inference and privacy-conserving on-device computation has been catalyzed by the journey towards these breakthroughs. Lightweight and energy-efficient architectures have been studied new with specific applications in embedded systems, in which cost-effectiveness versus trade-offs in accuracy and latency are the prevailing factors of technology Archana Reddy et al. [2]. As an example, binarized and quantized neural networks, resource-sharing processor architectures, and co-optimization of hardware and software, have made it possible to implement CNN-based gesture recognition systems on FPGAs and microcontrollers with remarkable performance and energy characteristics, and with state-of-the-art accuracy and operational performance even with severe resource constraints Zhang et al. [1]; Yu et al. [3]; Tsai et al. [4]; Nagaraju et al. [29]. These developments form the basis of gesture recognition being deployed in devices such as smartwatches that can control home appliances, the microcontroller-based robotic prostheses, and automation in the industrial setting Nguyen-Trong et al. [5]; Fu and Ye [6]; Kim et al. [7]. Vision-based systems have also been popularized because of its capability to give rich spatiotemporal details of the complex gestures, but it still faces practical limitations that include variations of lighting, occlusions, and hand shapes or skin color Tellaeche et al. [8]; Fu [9]. Deep CNN architectures, including residual and separable convolution, attention-based, and self-supervision, have continued to surpass traditional methods in gesture classification based on images and sensors, usually through transfer learning and data augmentation to improve the results on more difficult inputs Chen [10]; Patrona et al. [11]; Sasidharan et al. [12]. The need to have gesture recognition systems to effectively work on border devices has fast-tracked the study of quantization conscious training, pruning, model compression, and the deployment of specialized inference accelerators that lessen both memory footprint and computational activities. These optimizations allow edge AI systems to be executed in environments where constant access to cloud infrastructure is not convenient or where privacy issues require local inference exclusively. Gesture recognition in real time on systems like NVIDIA Jetson, Raspberry Pi, or specialized AI microcontrollers is now the norm, which shows that CNN-based models with pruning and tailoring can be made to approach the accuracy of a PC in HCI applications in the field Körösi et al. [13]. Several sensor modalities have been examined as well to enhance system robustness. An example is that radar-based gesture recognition is also usable in low-visibility environments and is less sensitive to background clutter compared to all-optical systems, and EMG- or IMU-based sensors detect the intent or movement of the muscles themselves by the body, which can be used in a variety of applications, including sign language recognition and physically impaired assistive systems as well as in controlling industrial robots Chmurski et al. [14]; Safa et al. [15]; Jaramillo et al. [16]; Lu et al. [17]. CNNs combined with sensor fusion, including multimodal deep learning, lead to systems that can adapt to users, device location, and environments to become more accessible to a wide range of users and with more practical reliability Muhoza et al. [18]. Yet, there are still obstacles, such as the ability to handle user-specific variations, resistance to adversarial situations, and the ability to reduce the computational cost, without compromising the recognition accuracy, that still drive the search of meta-learning, attention, and federated learning strategies tailored to edge HCI Fan et al. [20]. The interfaces between human and computers developed based on these technologies of gesture recognition are now transforming a wide area of fields of immersive interaction of reality, both virtual and augmented Lv et al. [21] to intelligent houses and helpful robotics that use gestures to provide natural user input Alabdullah et al. [19]; Anupama et al. [22]. Embedded systems use model compression and pruning, platform optimization (e.g. FPGA implementation), and data-specific data pipelines to get low-latency, high-throughput recognition and designers are increasingly using the hybrid CNN-RNN or spatiotemporal models to get gesture sequences that change with time Zhang et al. [23]. The recent developments in sign language recognition, traffic control, and even closed-loop control of industrial robotics prove once again the flexibility and influence of edge AI-enabled gesture recognition, and are the current trend in designing more inclusive, safe, and efficient HCI Dai et al. [24]; Rutishauser et al. [25]. It is interesting to note that supporting user privacy and data security is becoming an even more important consideration as these systems monitor subtle motions and behaviour, leading to a body of research in support of encrypted processing and locally constrained inference. The combination of EMG, radar, and vision signals with deep CNNs and their implementation on edge accelerators embody the range of engineering and algorithmic advances that combine to render HCI on a mass scale, in real time, and context-aware using gestures as a means of interaction. Scientometric, the rise of literature on edge-deployable CNN-based gesture recognition since 2018 can testify to the active development of the field, its involvement in cross-disciplinary interactions and its adherence to the tendencies in ubiquitous computing, wearable computing, and AI-driven interface Oudah et al. [26]; Jiang et al. [27]; Ramya et al. [30]. The quantitative trends show geometric growth in citations and publications, which means that gesture recognition systems have become and are becoming technically relevant, have impact on society, and are implemented at the base of next-generation human-computer interfaces Morshed et al. [28].

2 Methodology

The proposed study using a scientometric approach is an effort to systematically evaluate the intellectual framework and research dynamics of the gesture recognition systems with convolutional neural networks (CNNs) and embedded edge AI as the human computer interaction (HCI) interface. The Scopus and Web of Science databases were searched to retrieve data on the period between 2020 and 2025. A query in the form of a Boolean operator was created to locate the literature of interest, and the dataset was narrowed down by eliminating duplicates and abstracts, as well as documents in non-English languages, creating a curated corpus that can be used in the process of bibliometric evaluation. Scientometric analysis was performed with the help of VOS viewer and Biblioshiny (RStudio) to visualize collaboration networks and clusters of key words and thematic evolution. Measures like rate of growth of publications, citation effect, productivity of the authors, institutional contribution as well as international co-authorship trends were calculated. The networks of co-citation and co-authorship were plotted to discover the most active researchers and interaction centers. Keywords clustering also facilitated the identification of emerging research trends, such as lightweight architecture, multimodal fusion, and energy balanced models. By merging quantitative bibliometric measures and visualization through a network, the methodology can give a broad view of the developmental trend of the domain, the intellectual hubs, and the research direction in the future, thereby contributing to the evidence-based understanding of CNN-based gesture recognition and edge AI implementation to HCI.

Research Questions:

1. Examine publication trends in gesture recognition using CNNs and edge AI.
2. Identify leading authors, institutions, countries, and journals contributing to the domain.
3. Explore thematic clusters using keyword co-occurrence analysis.
4. Highlight emerging research areas and gaps that warrant further investigation.

Search Key

(TITLE-ABS-KEY ("Gesture Recognition") AND TITLE-ABS-KEY ("Convolutional Neural Networks") AND TITLE-ABS-KEY ("Human–Computer Interaction")) AND PUBYEAR > 2019 AND PUBYEAR < 2026 AND (LIMIT-TO (LANGUAGE, "English")).

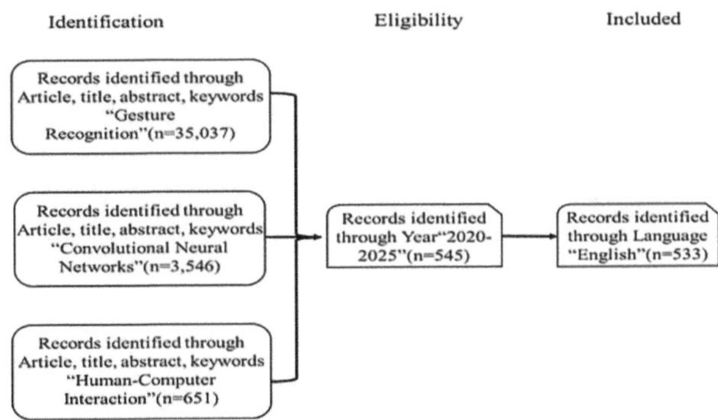

Fig. 1. Prisma Flow-Chart

As the Fig. 1 shows, about the study in gesture recognition and human-computer interaction, there was a scientific production of 2020–2025 every year. The trend indicates that the publications have been increasing steadily over time, which indicates the increment in academic interest in this field and progress. The trend shows that the role of such technologies as convolutional neural networks, deep learning, and edge AI is increasingly important to enhance gesture-based systems. The gradual increase in the amount of research papers is a testimony to the academic and industrial interest since the research papers are contributed by various authors and various institutions. Overall, the figure demonstrates the growing significance and the continuous tendency toward the rise of the research in the sphere of human-computer interaction and gesture recognition applications.

3 Results and Discussion

According to the scientometric review of 477 articles on gesture recognition and HCI published between 2020 and 2025, the rate of growth is 8.45/year, which means that this field is gradually developing. The collaboration intensity and average number of authors per paper and percent international co- authorship of 12.3 and 14.26 respectively makes there to be high global networking. According to the Law of Lotka analysis, about 80 percent of authors contribute to only one work which implies a broad but cut short research base. Bradford Law identifies IEEE Access, Sensors and IEEE Sensors Journal as the main sources of publication in the dissemination of the advancements. The works of Mitra (2007) and Oudah (2020) find their place in the list of seminal works cited as shaping the research path. On the country level, China and India lead the production, and South Korea, USA and Europe have an emerging presence, and it represents a varied but Asia-centric research ecosystem. In summary, keyword clustering focuses on CNNs, edge AI, and lightweight architecture as the existing hotspots, and privacy-preserving models and multimodal fusion are the way forward.

Fig. 2. Main Information

The Fig. 2 summarizes key bibliometric indicators for gesture recognition and human–computer interaction research between 2020 and 2025. The number of published documents is 477 and the number of sources is 161, with 2,605 authors. The field has an annual growth rate of 8.45 and there are no one-authored papers which indicate that there is a strong collaboration. There are 12.3 co-authors on average per document, and 14.26 percent of articles are co-authored internationally. These data cover 3,180 author keywords and 3,651 references, and the average age of documents is 2.21 years old. Johnson cited an average of 10.4 documents indicating the increased academic contribution of studies in this field.

Table 1 shows how many authors have accomplished their works in gesture recognition and human-computer interaction studies, according to the Law of Scientific Productivity as introduced by Lotka. The distributions reveal that nearly all of the authors published a single article, with 2,092 authors (80.3) having published a single article,

Table 1. Author's Productivity Through Lotka's Law

N. Articles	N. Authors	Freq
1	2092	0.80307102
2	268	0.10287908
3	76	0.02917466
4	48	0.0184261
5	17	0.00652591
6	9	0.00345489
7	20	0.00767754
8	8	0.00307102
9	4	0.00153551
10	8	0.00307102

which suggests there is a large, but scattered research community. The smaller shares of the authors are those with two or three papers (268 and 76 authors, respectively), but with smaller contribution to the overall output. There were a small number of very prolific authors who published more than five articles, less than 1% of the total. This biased distribution is in line with the inverse-square rule made by Lotka, according to which a small group of active scholars make most of the publications, which explains the collaborative and growing character of research in this field.

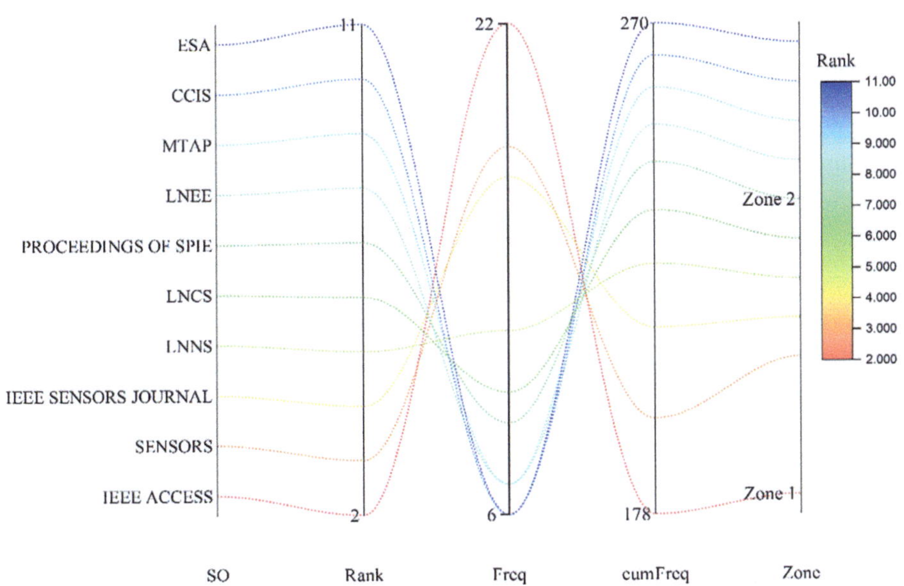

Fig. 3. Bradford's Law

The visualization of the Bradfords Law in Fig. 3 shows the distribution of the research sources in the field of gesture recognition and human-computer interaction. The plot in differentiating journals into productivity areas reveals that few sources provide most of the publications. IEEE access, sensors, and IEEE sensors Journal hold the position of Zone 1 of the list, as they are major publication sources in the field of research. Zone 2 is filled with LNCS, LNEE, MTAP journals and Proceedings of SPIE, which are journals of secondary, though, importance. The gradual curve between red and blue shows the variation in rank and frequency, which indicates the declining concentration of the papers by zones and reflects unequal but organized distribution of scientific output as is implied by the Law of Bradford.

Table 2 gives a closer description of the most impactful cited sources in the sphere of research on gesture recognition, with the number of citations provided. The entries provide the author(s), the title of the publication, the source, and the year of publication providing an exhaustive insight into the efforts of the major contributors to this field. The table outlines a large variety of works, such as surveys, reviews, and state-of-the-art deep learning methods used in recognizing hand gestures, processing sign languages, and human-computer interaction. Much-referenced literature, including the literature by Mitra, Rautaray and Oudah, highlights the background methods and technological development that have mostly influenced the research community. Other contributions that are worth noting are deep residual learning, the application of the convolutional neural network, and the newly emerging methods such as millimeter-wave radar-based gesture recognition. Summarizing these prominent references, as well as the effect of their citation, the table can be regarded as a useful tool in recognizing key trends, authors that influence the development of gestures recognition research, and its changing course.

Figure 4 depicts the distribution of research articles by major affiliations involved in the research in gesture recognition and human computer interaction. Dr. Vishwanath Karad MIT World Peace University is ahead with 27 publications, then the National Institute of Technology Silchar (25) and the University of Chinese Academy of Sciences (24). Others with significant contribution are Beijing Institute of Technology (22), southeast University (20) and Fudan university (19). There are several Chinese universities that demonstrate high attendance including Chongqing University of Posts and Telecommunications, Zhengzhou University and Hebei University of Technology. Visualization shows the institutional and global dissemination of the productivity of research in this field.

The network of co-occurring keywords and countries in gesture recognition and human-computer interaction studies in 2020–2025 is depicted in Fig. 5. Keywords like convolutional neural networks, gesture recognition and human computer interface are major clusters and are the hotspots of research. India is the dominant country in this sphere, which means that there is a high level of scholarly activity and partnerships. The time distribution indicates the steady change of the keywords with the time value in the sense of the increasing significance of deep learning and edge AI in the study of human-computer interaction. Visualization offers information regarding thematically developing and geographically contributing to this area of research.

Table 2. Most Local Cite References

CITED REFERENCES	CITATIONS
Mitra, Sushmita K., Gesture Recognition: A Survey, Ieee Transactions On Systems, Man And Cybernetics Part C: Applications And Reviews, 37, 3, Pp. 311–324, (2007)	18
Rautaray, Siddhath Swarup, Vision Based Hand Gesture Recognition For Human Computer Interaction: A Survey, Artificial Intelligence Review, 43, 1, Pp. 1–54, (2015)	17
Oudah, Munir, Hand Gesture Recognition Based On Computer Vision: A Review Of Techniques, Journal Of Imaging, 6, 8, (2020)	16
He, Kaiming, Deep Residual Learning For Image Recognition, Proceedings Of The Ieee Computer Society Conference On Computer Vision And Pattern Recognition, 2016-December, Pp. 770–778, (2016)	13
Sharma, Sakshi Aneja, Vision-Based Hand Gesture Recognition Using Deep Learning For The Interpretation Of Sign Language, Expert Systems With Applications, 182, (2021)	13
Atzori, Manfredo, Deep Learning With Convolutional Neural Networks Applied To Electromyography Data: A Resource For The Classification Of Movements For Prosthetic Hands, Frontiers In Neurorobotics, 10, Sep, (2016)	11
Molchanov, Pavlo A., Online Detection And Classification Of Dynamic Hand Gestures With Recurrent 3d Convolutional Neural Networks, Proceedings Of The Ieee Computer Society Conference On Computer Vision And Pattern Recognition, 2016-December, Pp. 4207–4215, (2016)	11
Cheok, Ming Jin, A Review Of Hand Gesture And Sign Language Recognition Techniques, International Journal Of Machine Learning And Cybernetics, 10, 1, Pp. 131–153, (2019)	10
Oyedotun, Oyebade Kayode, Deep Learning In Vision-Based Static Hand Gesture Recognition, Neural Computing And Applications, 28, 12, Pp. 3941–3951, (2017)	10
Chakraborty, Biplab Ketan, Review Of Constraints On Vision-Based Gesture Recognition For Human-Computer Interaction, Iet Computer Vision, 12, 1, Pp. 3–15, (2018)	9
Lien, Jaime, Soli: Ubiquitous Gesture Sensing With Millimeter Wave Radar, Acm Transactions On Graphics, 35, 4, (2016)	9

It is described in Fig. 6 as the country-wise distribution of publications in gesture recognition and human past interaction over the period 2020 to 2025. The list of China and India takes the leading positions, thus providing many publications, with South Korea, USA, and Germany in the following. The other involved contributors are Malaysia, Japan, Saudi Arabia, Italy, and Egypt, in which other countries like France, Spain, Turkey, and the United Arab Emirates contribute comparatively less but with significant contribution. The existence of very varied areas, such as Indonesia, Norway, Poland, and

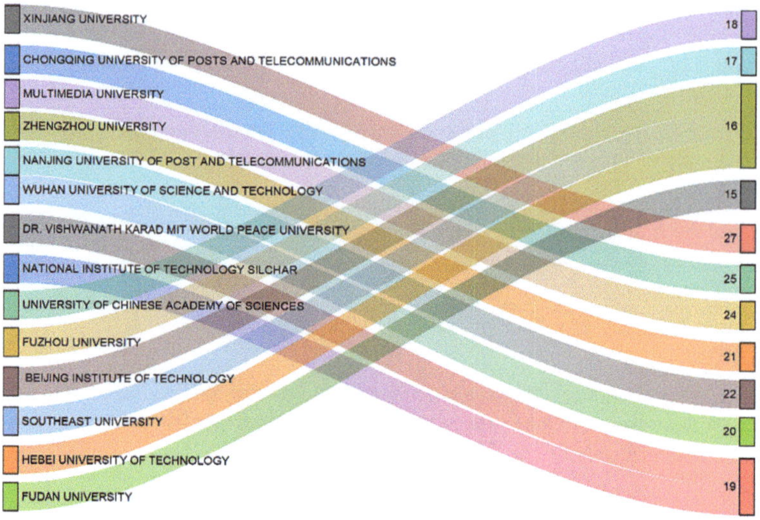

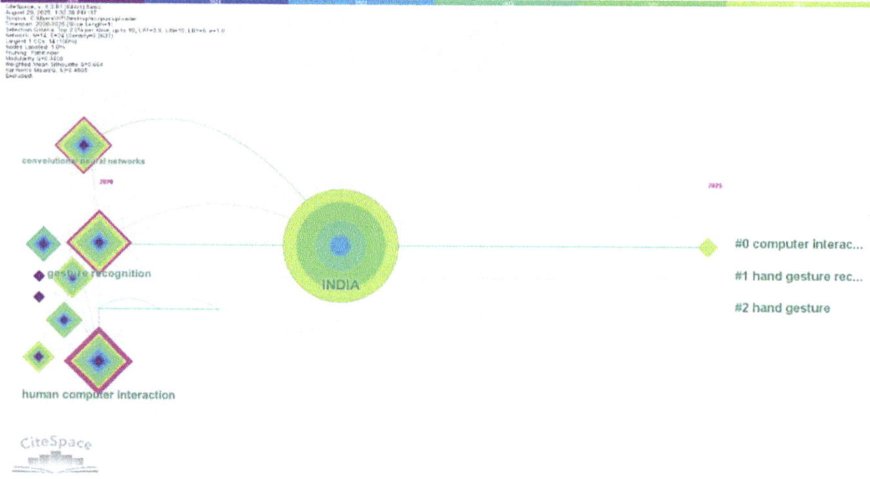

Fig. 4. Most Relevant Affiliations

Fig. 5. Keywords and Country Distribution in Human–Computer Interaction Research

Thailand, points to the global nature of the given research area, although Asian countries are evidently the ones that have the most publication rate and research activity.

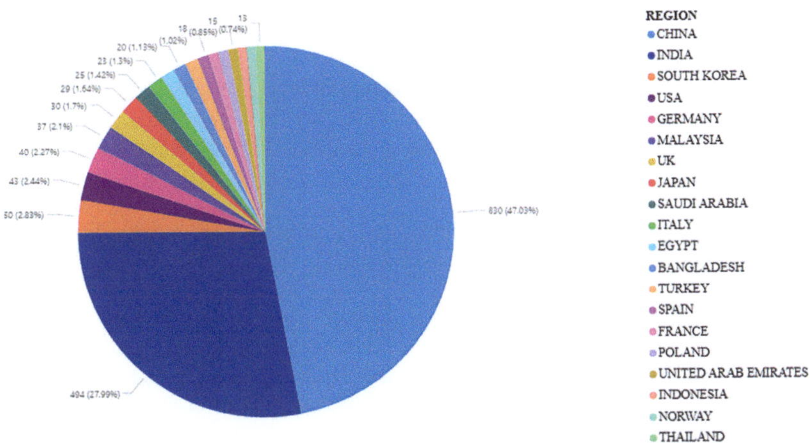

Fig. 6. Country Scientific Production

4 Conclusion

This paper is a scientometric review of gesture recognition systems that rely on convolutional neural networks (CNNs) and embedded edge AI, along with their development, current dynamics, and research contributions in the field of human-computer interaction (HCI). The discussion shows that CNNs have become the paradigm because of their high spatiotemporal features extraction power, robustness, and versatility in multimodal inputs. Simultaneously, increasing integration of edge AI highlights how the field is transitioning to real-time, low-latency, and privacy-preserving computation, and can be applied to wearables, robotics, healthcare and smart environments. Scientometric mapping shows that there is a gradual but consistent trend of growth in scholarly output all over the world with China, India, and the United States taking the lead. Network visualizations also reveal highly leveraged international partnerships, thematic communities around lightweight architectures, and broadening exploration of multimodal fusion and energy-efficient design. Even though a lot has been done, there are issues that cannot be resolved. Computational costs are high, power requirements are low, and they can withstand variability in the environment are barriers to large scale implementation. Moreover, the ethical matters, in particular, the privacy of users and the safe data processing must be discussed in more detail to ensure that it is addressed in a credible way. The research should thus focus on future studies to focus on lightweight CNN architectures to be used in constrained hardware, cross sensor fusion techniques to enhance robustness and privacy preserving algorithms to be deployed at the edges. This research synthesized scientometric evidence with technical knowledge to map the intellectual structure of gesture recognition research, as well as to offer an evidence-based roadmap to specify next-generation HCI. Intersection of CNNs and embedded edge AI is a key facilitator of intelligent, adaptable, and human-centric interaction technologies that are bound to revolutionize digital ecosystems across the world.

References

1. Zhang, Y., et al.: An energy-efficient BNN accelerator with two-stage value prediction for sparse-edge gesture recognition. IEEE Trans. Circ. Syst. I, Regular Papers: A Publication of the IEEE Circuits and Systems Society, pp. 1–14 (2023). https://doi.org/10.1109/tcsi.2023.3320175

2. Archana Reddy, R., Al-Attabi, K., Chandana, B., Hatem Mohammed, I., Bhuvaneswary: Prediction of stock market price using bi-directional recurrent neural network. In: 2023 International Conference on Ambient Intelligence, Knowledge Informatics and Industrial Electronics (AIKIIE), pp. 1–5 (2023)

3. Yu, A., Qian, C., Guo, Y.: A real-time hand gesture recognition system on raspberry pi: A deep learning-based approach. In: 2024 IEEE 21st Consumer Communications & Networking Conference (CCNC), pp. 499–506. https://doi.org/10.1109/CCNC51664.2024.10454652

4. Tsai, T.-H., Ho, Y.-C., Chi, P.-T.: Hardware architecture design for hand gesture recognition system on FPGA. IEEE Access: Pract. Innovat. Open Solut. **11**, 51767–51776 (2023). https://doi.org/10.1109/access.2023.3277857

5. Nguyen-Trong, K., Vu, H.N., Trung, N.N., Pham, C.: Gesture recognition using wearable sensors with bi-long short-term memory convolutional neural networks. IEEE Sens. J. **21**(13), 15065–15079 (2021). https://doi.org/10.1109/jsen.2021.3074642

6. Fu, Z., Ye, W.: A 593nJ/inference DVS hand gesture recognition processor embedded with reconfigurable multiple constant multiplication technique. IEEE Trans. Circ. Syst. I, Regular Papers: A Publication of the IEEE Circuits and Systems Society, **71**(6), 2749–2759 (2024). https://doi.org/10.1109/tcsi.2024.3387998

7. Kim, E., Shin, J., Kwon, Y., Park, B.: EMG-based dynamic hand gesture recognition using edge AI for human–robot interaction. Electronics **12**(7), 1541 (2023). https://doi.org/10.3390/electronics12071541

8. Tellaeche Iglesias, A., Fidalgo Astorquia, I., Vázquez Gómez, J.I., Saikia, S.: Gesture-based human machine interaction using RCNNs in limited computation power devices. Sensors (Basel, Switzerland) **21**(24), 8202 (2021). https://doi.org/10.3390/s21248202

9. Fu, D.: Human-computer interaction gesture control recognition for high-performance embedded AI computing platform. Adv. Comput. Commun. **4**(3), 210–214 (2023). https://doi.org/10.26855/acc.2023.06.021

10. Chen, Y.: Design and application of hand gesture recognition in embedded image processing system using modified convolutional neural network. In: 2025 3rd International Conference on Data Science and Information System (ICDSIS), pp. 1–7 (2025). https://doi.org/10.1109/ICDSIS65355.2025.11070688

11. Patrona, F., Mademlis, I., Pitas, I.: Self-supervised convolutional neural networks for fast gesture recognition in human-robot interaction. In: 2021 10th International Conference on Information and Automation for Sustainability (ICIAfS) (2021). https://doi.org/10.1109/ICIAfS52090.2021.9605915

12. Sasidharan, R., Varghese, A., Rashmi, M., Surendran, S.P.: An efficient attention model with Critical Frames Identification for sign language recognition (CRAM-SLR). J. Comput. Cogn. Eng. (2025). https://doi.org/10.47852/bonviewjcce52025288

13. Körösi, L., Kajan, S., Kováč, M., Skirkanč, J., Sekaj, I.: Real-time servo control with gesture recognition on nvidia Jetson using CNNs. Cybern. Inform. **2025**, 1–6 (2025)

14. Chmurski, M., Mauro, G., Santra, A., Zubert, M., Dagasan, G.: Highly optimized radar-based gesture recognition system with Depthwise Expansion Module. Sensors (Basel, Switzerland) **21**(21), 7298 (2021). https://doi.org/10.3390/s21217298

15. Safa, A., et al.: Improving the accuracy of spiking neural networks for radar gesture recognition through preprocessing. IEEE Trans. Neural Netw. Learn. Syst. **34**(6), 2869–2881 (2023). https://doi.org/10.1109/TNNLS.2021.3109958

16. Jaramillo-Yánez, A., Benalcázar, M.E., Mena-Maldonado, E.: Real-time Hand Gesture Recognition using surface electromyography and machine learning: a systematic literature review. Sensors (Basel, Switzerland) **20**(9), 2467 (2020). https://doi.org/10.3390/s20092467

17. Lu, J., Liang, P., Rhim, J.C., Zhang, X., Qin, Z.: Efficient convolutional neural network for real-time EMG pattern recognition system on edge devices. In: 2023 11th International IEEE/EMBS Conference on Neural Engineering (NER), pp. 1–5 (2023). https://doi.org/10.1109/NER52421.2023.10123741

18. Muhoza, A.C., Bergeret, E., Brdys, C., Gary, F.: Multi-position human activity recognition using a multi-modal deep convolutional neural network. In: 2023 8th International Conference on Smart and Sustainable Technologies (SpliTech) (2023). https://doi.org/10.23919/SpliTech58164.2023.10193600

19. Alabdullah, B.I., et al.: Smart home automation-based hand gesture recognition using feature fusion and recurrent neural network. Sensors (Basel, Switzerland) **23**(17) (2023). https://doi.org/10.3390/s23177523

20. Fan, X., Zou, L., Liu, Z., He, Y., Zou, L., Chi, R.: CSAC-Net: fast adaptive sEMG recognition through attention convolution network and model-agnostic meta-learning. Sensors (Basel, Switzerland) **22**(10), 3661 (2022). https://doi.org/10.3390/s22103661

21. Lv, Z., Poiesi, F., Dong, Q., Lloret, J., Song, H.: Deep learning for intelligent Human-Computer Interaction. Appl. Sci. (Basel, Switzerland) **12**(22), 11457 (2022). https://doi.org/10.3390/app122211457

22. Anupama A, Yamikar, R.S., Renuka K.M.: AI-driven gesture control for industrial robots: a vision-based approach for enhancing human-robot collaboration. World J. Adv. Res. Rev. **20**(2), 1498–1506 (2023). https://doi.org/10.30574/wjarr.2023.20.2.2254

23. Zhang, Y., Jiang, X.: Recent advances on deep learning for sign language recognition. Comput. Model. Eng. Sci.: CMES **139**(3), 2399–2450 (2024). https://doi.org/10.32604/cmes.2023.045731

24. Dai, F., Yi, X., Chai, S.: Hand gesture recognition for traffic signals based on improved lightweight convolutional neural network. In: 2023 2nd International Conference on Artificial Intelligence and Computer Information Technology (AICIT) (2023). https://doi.org/10.1109/AICIT59054.2023.10278005

25. Rutishauser, G., Hunziker, R., Di Mauro, A., Bian, S., Benini, L., Magno, M.: ColibriES: A milliwatts RISC-V based embedded system leveraging neuromorphic and neural networks hardware accelerators for low-latency closed-loop control applications. In: 2023 IEEE International Symposium on Circuits and Systems (ISCAS), pp. 1–5 (2023). https://doi.org/10.1109/ISCAS46773.2023.10181726

26. Oudah, M., Al-Naji, A., Chahl, J.: Hand gesture recognition based on computer vision: a review of techniques. J. Imag. **6**(8), 73 (2020). https://doi.org/10.3390/jimaging6080073

27. Jiang, S., Kang, P., Song, X., Lo, B., Shull, P.: Emerging wearable interfaces and algorithms for hand gesture recognition: a survey. IEEE Rev. Biomed. Eng. **15**, 85–102 (2022). https://doi.org/10.1109/RBME.2021.3078190

28. Morshed, M.G., Sultana, T., Alam, A., Lee, Y.-K.: Human action recognition: a taxonomy-based survey, updates, and opportunities. Sensors (Basel, Switzerland) **23**(4) (2023). https://doi.org/10.3390/s23042182

29. Nagaraju, T.V., Gobinath, R., Awoyera, P., Abdy Sayyed, M.A.H.: Prediction of California bearing ratio of subgrade soils using artificial neural network principles. In: Communication and Intelligent Systems, pp. 133–146. Springer Singapore (2021)

30. Ramya, U., Sammaiah, P., Subbarao, M., Mahesh, V., Satheesh Kumar, K.: Advancements in carbon fiber reinforced liquid silicone rubber: a scientometric study. J. Inst. Eng. (India) Series C (2024). https://doi.org/10.1007/s40032-024-01146-9
31. Shyamala, G., Gobinath, R., Sarla, P., Shewale, M.: Analysis of compressive strength characteristics of mineral admixture in concrete containing various gelled materials using Artificial Neural Networks. IOP Conf. Ser. Mater. Sci. Eng. **981**(3), 032093 (2020). https://doi.org/10.1088/1757-899x/981/3/032093

Open Access This chapter is licensed under the terms of the Creative Commons Attribution-NonCommercial-NoDerivatives 4.0 International License (http://creativecommons.org/licenses/by-nc-nd/4.0/), which permits any noncommercial use, sharing, distribution and reproduction in any medium or format, as long as you give appropriate credit to the original author(s) and the source, provide a link to the Creative Commons license and indicate if you modified the licensed material. You do not have permission under this license to share adapted material derived from this chapter or parts of it.

The images or other third party material in this chapter are included in the chapter's Creative Commons license, unless indicated otherwise in a credit line to the material. If material is not included in the chapter's Creative Commons license and your intended use is not permitted by statutory regulation or exceeds the permitted use, you will need to obtain permission directly from the copyright holder.

AI-Based Virtual Interviewer System Using NLP and Emotion Detection

J. M. Bershika[✉] and Golden Nancy

Division of Artificial Intelligence and Machine Learning, Karunya Institute of Technology and Sciences, Coimbatore, Tamil Nadu, India
bershikaj@gmail.com

Abstract. The growing use of remote and mass-recruitment tactics has led to the realization of a demand in the intelligent and impartial interview systems that can screen the candidates out of the conventional resume screening. In this paper, the Author describes an AI-based Virtual Interviewer System which combines with natural language processing, multimodal emotion recognition, and adaptive question generation to facilitate automated evaluation of candidates. The system allows the applicants to post their resumes, schedule interviews and engage in virtual interview in real time using a web-based interface created with Flask. Semantic similarity algorithms between resumes and job descriptions are used to conduct resume parsing and candidate-role matching based on transformer-based NLP models, which are accessed via the Gemini API. In interviews, speech of candidates is transcribed and analyzed according to their relevance, fluency and accuracy in the domain, whereas facial expressions are recognized with the Deep-Face to detect emotional cues, including confidence and engagement. Depending on real time performance, the system is dynamic in adapting interview questions to probe technical and behavioral stability. Weighted fusion model is a combination of linguistic, emotional, and contextual measures that create structural scorecards to recruiters. Human ratings, low latency and scalable performance under concurrent interview loads are all shown to be well correlated with experimental evaluation. The suggested system allows increasing efficiency, fairness, and transparency in the recruitment process, which presents a feasible answer to the contemporary hiring setting.

Keywords: Automated interview · AI based recruitment · resume parsing · video analysis · emotion detecting · virtual interviewer · NLP · DeepFace · Gemini API · interview automation

1 Introduction

The high rate of digitalization of industries has transformed the recruitment process undertaken by traditional face to face interview to smart, computerized, and online tests. This trend implies that as the level of remote work and employee search in different parts of the world increases, there is the necessity to provide organizations with scalability, objectivity, and equitability in judging the candidates. The technologies of Artificial

© The Author(s) 2026
J. C. Bansal et al. (Eds.): SCIS 2025, LNNS 1929, pp. 69–85, 2026.
https://doi.org/10.1007/978-3-032-22911-3_6

Intelligence (AI) and, to be more precise, Natural Language Processing (NLP) and emotion detection became a potentially efficient alternative to make the decision-making process more efficient by combining the linguistic and behavioral data.

The conventional interviews will probably be constrained with few shortcomings such as interviewer bias, inconsistency of evaluation, and inability to scale when large pools of candidates are used. Similarly, the existing AI-based recruitment tools are likely to be partial in their offerings some can read and rank the resumes, others can do a sentiment or facial expression analysis, and others are powered by the predetermined questionnaire that is not altered according to the response provided by the job seeker. These fragmented approaches do not offer some comprehensive system capable of performing the assessment of the technical skills and behavioral traits in real-time.

The proposed AI-Based Virtual Interviewer System addresses the following gaps in the identical platform through the use of NLP, emotion detection, adaptive questioning, and automated scheduling. The system is used to evaluate the applicants in terms of their behavioural and technical aptitudes and the facial emotion identification through the Gemini API to read resumes and perform a linguistic analysis and Deep Face to identify facial emotion in real-time. It also enhances its personalization and efficacy through the inclusion of dynamic and domain specific question generation and automated scheduling and HR dashboard provides organized scorecards and evidence based information to recruiters. All these abilities combined enhance the system to be fair, transparent and flexible in decision making concerning recruitment.

1.1 Problem Statement

The key issue that was resolved in this article is that no detailed and scalable virtual interview system was developed that was capable of integrating natural language understanding, multimodal emotion detection, adaptive question formulation, and automatic time management. It is aimed to create and launch an AI-based platform, which will be less biased, consistent, and will offer a fair yet strict selection of candidates by position and discipline.

1.2 Contributions of the Paper

The significant contributions of the research are as follows:

Implementation and design of an end-to-end AI virtual interviewer that combines resume parsing, real-time interview analysis and automated scheduling in a single platform.

Creation of a multimodal candidate assessment system based on natural language understanding, speech analysis and facial emotion recognition.

Implementation of an adaptive question generation system to dynamically change the direction of the interview in regards to responses by candidates.

Development of a weighted fused fashion of scoring model that balances technical accuracy and behavioral understanding and yet, is fair.

Scalability, low latency, and high correlation with human recruiter judgments: experimental validation.

2 Literature Survey

A block of an AI-controlled virtual interviewer proposed by Shekar et al. [1] aimed to automatize the process of interviews performed through the use of conversational AI approach. Their system demonstrated that it was possible to replace the initial interviews with humans with the interaction on the basis of AI which would reduce the work of recruiters. However, the analysis was not conducted in detail of behavior or multimodal analysis because the focus of the study was primarily on dialogue automation. On the other hand, the proposed system is not confined to a conversation flow as it involves both assessment of reactions based on the results of NLP-based data and the analysis of emotions and confidence, which will assist in making the evaluation of the candidates more extensive and more evidenced-based.

Pathaka et al. [2] explored the issue of how a person can apply computer vision techniques in the analysis of interviewees in AI-based video interviews, non-verbal communication, such as facial expression and posture. Even though their work was feasible to demonstrate the significance of visual clues in the analysis of candidate behavior, it relied primarily on visual aspects and the slightest on the content of language. The proposed system would resolve this limitation of taking into account the responses and behavioral indicators jointly to ensure that the technical competence and the quality of communication are taken into account alongside non-verbal messages.

Lokhande et al. [3] recommended AI-driven mock interview behavior recognition that is focused on producing feedback regarding the candidates. Their approach was useful both in regard to post-interview feedback and inefficient in realtime flexibility and full semantic interpretation of candidate answers. In comparison, the proposed framework has dynamic NLP analysis and responsive questioning in which flow of interview and level of assessment can vary dynamically with the performance of the candidates.

Bajaj [4] designed an emotion and confidence recommender of mock interviews according to artificial intelligence and the importance of the affective states in interview performance. Although the classifier was effective in identifying emotional cues, it was more of a model which was not part of an interview pipeline. The proposed system will be a refinement of this as it will combine emotion and confidence analysis into an end-to-end virtual interviewer which will be a mixture of resume parsing, execution of the interview, scoring, and reporting.

Umaraniya and Das [5] have suggested utilizing the AI interview mocker which emphasizes the emotion and confidence category. Their analysis revealed that affective nature could add value to the quality of the interview feedback but not along with resume background and technical response evaluation. The proposed solution will fill this deficiency as emotional indicators will be placed in perspective of linguistic and role specific relevance and expectations which will translate to fairer and more equal ratings of the candidates.

Kulkarni et al. [6], presented a mock interview assessor that would be AI-powered that would assist in the automation of the scoring of the candidates. Though the system had shown possibilities in standardization of the evaluations, it was founded on preset evaluation criteria and there was no flexibility on the direction the interview took. Nonetheless, the suggested system suggests adaptive generation of questions, which is informed by similarity scores on NLP and would enable personalized and role-conscious interviews to explore strengths and weaknesses of the candidates in a more effective way.

The AI-based mock interview assessor developed by Golande et al. [7] is grounded on scoring performance and feedback. Although their system was more objective than manual interviews, it did not take into account multimodal fusion and scalability. The proposed structure is the extension of the provided work in terms of the introduction of multimodal behavioral analysis, real-time processing and scalable architecture that allows it to be used in the context of large-scale recruitment applications.

Tajik [8] examined the idea of introducing emotional intelligence based on AI in learning language systems to promote speaking competence in an adaptive feedback system. Even though recruitment was not the particular goal of the study, it could point to the improvement in communication by real-time emotional feedback. The system transforms this concept to the interview context where emotional intelligence along with the evaluation of the applicants are employed to gauge confidence, interest and an understanding of the interview.

Sylvara et al. [9] also examined how personality-based employment interviews could be automated through the assistance of an AI chatbot and proved that it was a productive method of evaluating personality traits. Despite its virtues, the model was largely conversational cues-based and could not be behaviorally verified in a multimodal manner. The provided system adds personality inference to explicit linguistic scoring and emotion-aware analytics which enhances the degree of reliability and reduces the risks of misidentifying.

Malk and Diwan [10] gave a broad overview of the emotion detection in speech, which included the trends, challenges, and future directions. The issues that were raised in their analysis were that, it has variability in accent and sensitive to noise. The above insights form the basis of the given system, in which emotional cues delivered by speech are supplemented by analyzing text and images, thereby eradicating the drawbacks of emotion detection through single-modality.

Nithya [11] proposed a sharp job interview preparation and career promotion plan where the emphasis does not lie on evaluation of applicants but rather, it is guided towards applicant training. The system was a good preparation instrument, however, it was not recruiter-facing and objective-oriented. The proposed framework is distinct because it will involve automated candidate evaluation and recommendation to the recruiters, but will also give the candidates systematic feedback.

As per the review, Al-Azani and El-Alfy [12] explored the multimodal data of emotional AI by emphasizing the necessity to diversify data and fuse it with the help of fusion methods. Their effort discovered problems of prejudice and generalization with regard to demographics. The above findings were incorporated into the proposed system of the fairness-conscious evaluation and transparency tools whereby the equality of safety depends on gender differences, accent differences, and behavioral differences of an individual.

Okochi et al. [13] developed a computer human with human emotional perception to communicate through human computer interface and it was found more interactive with the user because of its emotion sensitive responses. The system was not created though to be tested and graded. The proposed virtual interviewer will elaborate on the emotional cognizance to measurable interview analytics, where the affective cues are translated to decipherable performance metrics.

Prince et al. [14] have proposed a multimodal AI system to enhance non-verbal communications when delivering a speech to the crowd. Their consequence determined the topicality of the study of gesture and expression, but which was directed towards training rather than measurement. The proposed system repeats the similar multimodal understanding to recruitment and applies it to the linguistic analysis to determine the suitability of the candidate in interview contexts.

Eloqify by Nikam [15] presented a smart interview partner that assists the candidates during the interviews. Though it is a beneficial supporting tool, it was not objectively rated and recruiter-oriented statistics. The suggested system does not just help but provideth a structured assessment, comparative ratings, and support of decision making and can therefore be used in institutional recruitment.

Novelty of the Proposed Work:

According to the literature, it is evident that the existing solutions revolve around non-integrated aspects of automating the interview such as conversational agent, emotion detection or mock interview feedback. The proposed system has been unique in that it integrates resume parsing, NLP based response analysis, multimodal behavior analysis, adaptive question generation and recruiter-facing analytics into a single end-to-end system. It is further concerned with fairness, transparency, scalability, and real-time adaptability, this is in comparison with the past, where the evaluation of candidates could have been favorable and consistent. This AI-enabled understanding is a massive enhancement to the existing AI-driven methods of interviews and it is a sound base that AI-aided, ethical and scalable automated recruitment can thrive.

3   Proposed Methodology

The proposed Virtual Interviewer System is the combination of Natural Language Processing (NLP), multimodal emotion recognition, adaptive questions generation, automated scheduling, and recruiter-facing dashboards. This integrative design is effective, fair and customized in the aspects of recruitment since it is a mix of linguistic testing, behavioral study in addition to contextual flexibility to form a composite image of the candidate. Figure 1 demonstrates the overall structure that allows seeing how the client interface, application layer, AI models, and HR dashboard are interconnected.

3.1   Candidate Resume Parsing and Onboarding

The applicants begin with the process by using a web application based on Flask and uploading their resumes and providing the required details including their skills and skills, education, past experience, and preferences related to their roles. The Gemini API also re-parses with similarity models using the name entity recognizer and in-built models and retrieves the structured data of competencies, qualification and work history. The semantic similarity measures are applied to compare the use of the attributes that have been parsed and the presence of the job role profile. The successful match

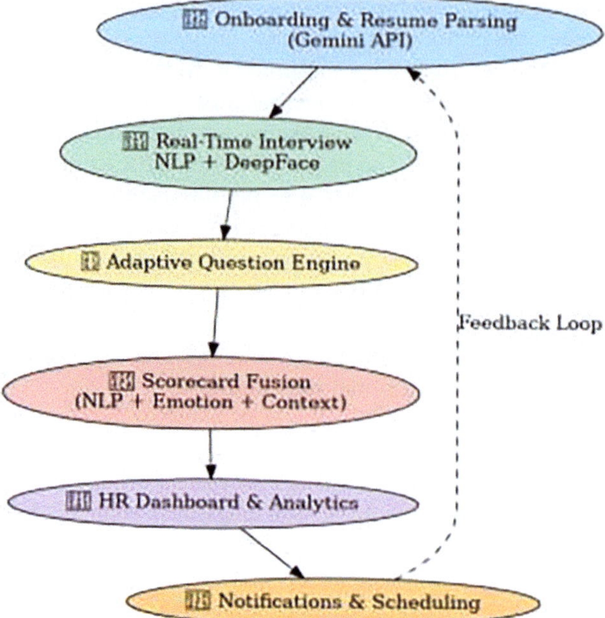

Fig. 1. Architecture Diagram

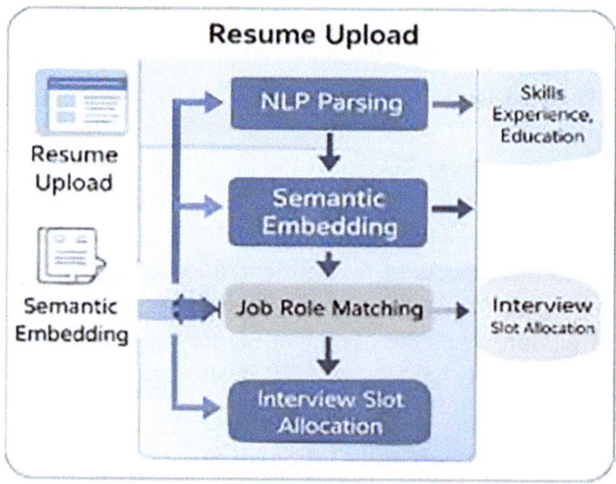

Fig. 2. Resume Parsing & Onboarding Workflow

will automatically show the interview slots which shall be sent through the integrated communication APIs to the candidates over secure email or SMS.

As shown in Fig. 2, semantic resume analysis enables efficient and unbiased candidate shortlisting prior to interview initiation.

3.2 Multimodal Analysis and Real Time Interview

The audio and video stream of the interview is recorded in the web-based interface. ASR is an automatic speech recognition system that converts speech into text and is evaluated by Gemini NLP models in terms of technical accuracy, domain and linguistic accuracy. At the same time, DeepFace does the video stream analysis that is used to diagnose such live face emotions as confidence, nervousness and attentiveness. To enhance reliability, preprocessing steps such as face alignment and histogram normalization are applied and temporal smoothing is applied to make combinations between frames and predictions. The combination of textual and emotional signals creates a whole behavioral image, so that the evaluation of behavior of candidates is not only capable of infusing both non and verbal elements of performance in the candidate.

3.3 Content Generation and Individualization

The adaptive questioning engine is one of the greatest innovations of the system. Question-generation models, which are trained on data specific to a domain, modify the flow of a certain interview based on the performance of the applicant. As an example, a semantic similarity score of the technical response of a candidate is low; then a focused follow-up question is designed, otherwise, the system replaces with behavioral or situational questions. This versatile thinking is in order to ensure that interviews can be tailored to fit, explore weak points further and sufficiently analyze problem solving competencies.

3.4 Comprehensive Review and Generation of Scorecards

The system will then compile the data after the session in an organized scorecard. Components include:

Content relevance, fluency, domain accuracy (NLP Analysis Scores).

Speech Qualitative Measures (clarity, speed)

Emotions Indicators (stability, engagement, level of confidence)

Contextual Metrics (consistency, adjustability during follow-ups)

The resulting candidate evaluation score is calculated by weighted fusion model as indicated in Eq. (1):

$$FinalScore = 0.75 NLP\_Score + 0.15 Emotion\_Score + 0.10\, Contextual\_Score$$

$$(1)$$

This weighting will make certain that both technical and linguistic accuracy will be the main criteria of evaluation, but the behavioral insights will support the evaluation and will not prevail. Through the HR dashboard, the recruiters are provided with instant feedback.

3.5 HR Dashboard and Analytics

Recruiter facing dashboard gives a single-point perspective of the candidate transcripts, interview recordings and analytics. It helps to filter, rank, and compare the candidates side-by-side to make a hiring decision based on data. Such tools as visual analytics display the metrics, which include average scores, behavioral stability graphs, and performance trends. Snippets of evidence, e.g. flagged quotes or emotional shifts are provided to enhance transparency and accountability in decision-making.

3.6 Scheduling and Management of Notifications Automation

The scheduling module is used to automate the process of coordinating the interviews between initiation and finalising. After shortlisting a candidate, available slots are scheduled against interviewer schedules using built-in schedules. The services of notification and reminders are provided through secure email and SMS, so that candidates could verify, reschedule, or cancel interviews. So by having automated reminders, no-shows are reduced and real-time synchronization of both interviewer and interviewer calendars to keep time under control.

3.7 Scalability, Security and Bias Mitigation

The system architecture is made to be scalable horizontally, where it supports parallel interviews by using containerized microservices and inference with GPUs. Data feeds are encrypted with TLS protocols, and the data stored is encrypted with AES protocols to abide by the privacy rules. Consent of the candidates is taken prior to every interview. The mitigation strategies are balanced training data, fairness-sensitive scoring, and frequent auditing of NLP and emotion recognition models to ensure gender, cultural, and accent biasness are addressed.

3.8 Limitations

Although the system has its benefits, there are some limitations to the system. Good and reliable performance is needed, whereby good audio and video input is needed; extreme background noise or bad lighting can reduce accuracy. The current implementation is mainly enacted to support English and a small number of popular languages, and only a small amount of multilingual is supported. Besides, real time emotion detection is computationally expensive and may need gpu resources that increase the cost of operation when it is used at large scale.

3.9 Ethical Considerations

The key principles that were placed in the system design were equity, privacy and transparency. Training data were gender balanced, there were cultural and linguistic differences to reduce bias and evaluation involved controls against uncertainty were factored in evaluation. The data of candidates was processed with explicit consent and encrypted on the storage level and during their transfer through the TLS protocols, and it adhered to the type of compliance of GDPR. We did not use black-box decision-making by providing transparency in which we provided snippets of evidence, and an apparent explanation of why the scores were the way they were. This is needed to ensure that efficiency gains as a result of automation are balanced with responsibility, trust and respect to the individual rights in the recruitment situations.

4   Results and Discussion

4.1   NLP Scoring Reliability

The natural language analysis module showed a large degree of correspondence to human ratings. Consensus on the expert raters was at a Cohens Kappa of 0.78 and correlation with the rubric based scores was 0.82. Accuracy, precision and recall were 89.2, 88.5 and 87.9, respectively, with ROC-AUC of 0.91. In order to have consistent benchmarks, inter-rater agreement was assessed on three human raters, yielding a Fleiss Kappa of 0.81, verifying the consistency of ground-truth labels. Semantic similarity models that were integrated into a rubric-based structure were used to mitigate personal bias whereas role-relevant assessment was facilitated by the use of skill-specific ontologies.

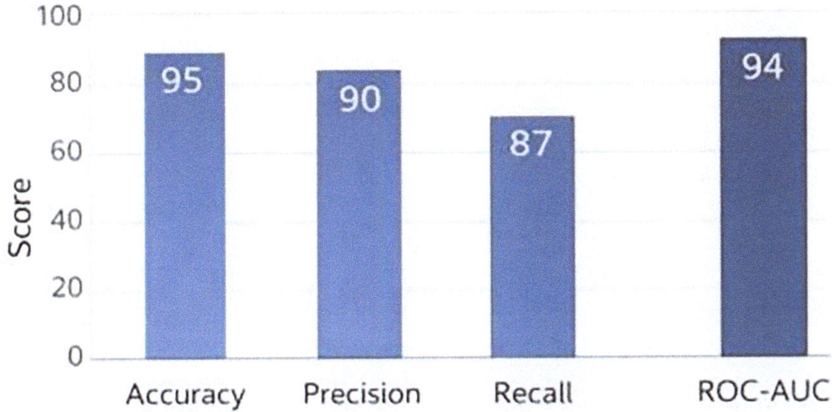

Fig. 3. NLP Scoring Performance

Measurements of evaluation of the NLP module that demonstrated good agreement with the human judgments. Through NLP scoring module as indicated in Fig. 3, the module exhibits high consistency with human judges in terms of accuracy, precision, recall, and ROC-AUC measure.

4.2   Accuracy of Detection of Emotions

The analysis of emotions provided promising results, with audio-based detection having a macro-F1 score of 0.78 and visual-based detection having 0.71. Accuracy was 82% in weighted accuracy in both modalities. How often should ROC-AUC reached was 0.84 speech and 0.79 facial cues. A temporal smoothing method was used to eliminate short-lived variations and to ensure the predictions were not volatile, it cut the volatility by approximately 12%.

Table 1, provides the summary of the performance of emotion detection module on audio and visual modalities, with the strength of speech-based cues being superior in comparison to facial expressions.

Table 1. Emotion Detection Performance

Modality	Macro-F1	Accuracy	ROC-AUC	Key Detected States
Audio-based	0.78	84%	0.84	Neutrality, Joy, Concern, Anger, Sadness
Visual-based	0.71	79%	0.79	Neutrality, Joy, Concern, Anger, Sadness

findings point to the fact that vocalic characteristics were typically more reliable than visual ones. However, the two channels were efficient to capture the key emotional categories.

4.3 Fusion Scoring Effectiveness

The fusion system took into account language, emotion, and contextual inputs with 75, 15 and 10 weight respectively. By removing emotion detection, the correlation with expert judgments declined by 8% and by removing adaptive questioning, the fairness ratings declined by 6%. The overall fusion model had an F1-score of 0.87, which is compared to 0.81 of the text only-model and 0.73 of the emotion only-model.

In order to test consistency, a five-fold cross-validation experiment was carried out. The mean F1-score was 0.86 plus or minus 0.02 that was robust between validation splits.

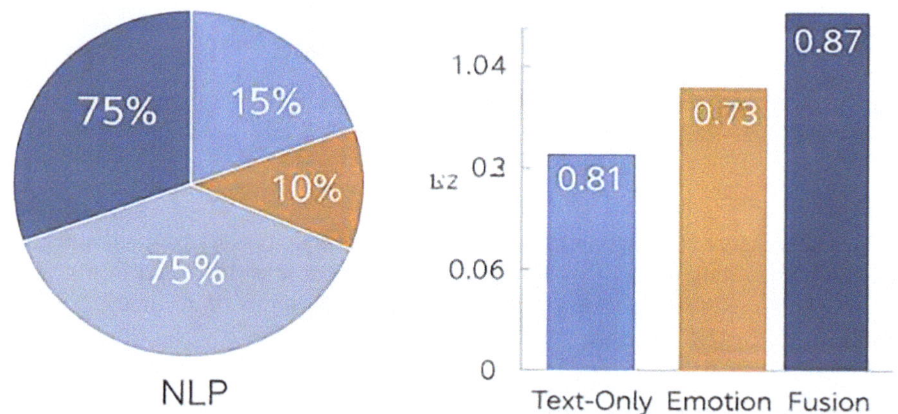

Fig. 4. Fusion Scoring and Ablation

The analysis of ablation as shown in Fig. 4 shows clearly the input of each evaluation module to the overall performance of the system. The exclusion of the emotion analysis element leads to a significant drop in correlation with the expert judgments and the exclusion of adaptive questioning yields lower fairness scores. These findings verify that the weighted fusion strategy that is suggested proves to be effective in balancing linguistic and behavioral knowledge to produce stronger and more equitable candidate ratings.

4.4   Scalability and Real-Time Interaction

With a constant mean latency of 2.3 s per candidate turn, the system was below the 2.5-s limit needed to have natural interaction. It was able to sustain well above 50 simultaneous interviews with no significant slowdowns using auto-scaling. Stress testing ensured that it was operational at 100 concurrent sessions without increasing latency, then gradually the latency started to increase. The use of GPUs was also found to average 72 meanwhile the use of memory was stable at 65 which demonstrated that the framework allocated workload effectively.

Table 2. Real-Time Interaction Metrics

Metric	Value	Target	Notes
Avg Latency/Turn	2.3 s	$\leq$2.5 s	Achieved through optimized ASR + GPU usage
Concurrent Capacity	50 + sessions	50	Auto-scaled using Kubernetes
Throughput	~200/hr	–	Measured on 4-core GPU node
Resource Usage	72% GPU, 65% RAM	<80%	Stable under heavy load

Table 2 shows real-time performance metrics of the system to show that the proposed architecture is able to operate within the latency and scalability constraints of live interviews.

Resource Comparison: When using GPU-backed deployment (NVIDIA T4), the response time remained to 2.3s, whereas CPU-only inference raised the average latency to 4.9s. An average of 0.09 per interview as proposed by cost profiling indicated that the strategy could be used at large scale.

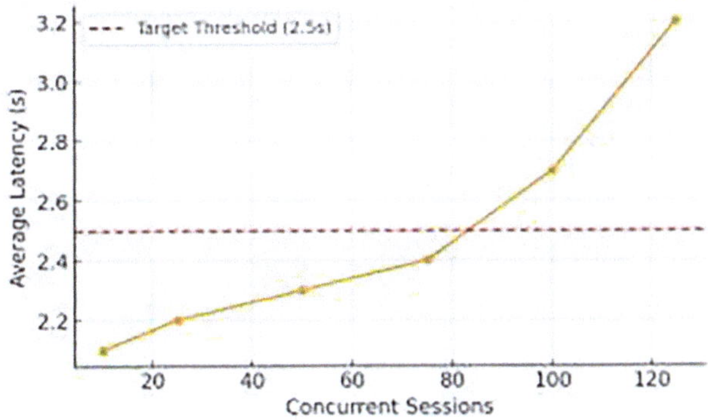

Fig. 5. Scalability Graph – Latency vs. concurrent sessions.

Figure 5 demonstrates that system latency remains within acceptable limits up to 50 concurrent interview sessions, validating real-time performance.

5   Error Analysis

Despite the generally high level of reliability of the framework, some difficulties were noted:

In dark settings, there were frequent lighting issues that caused a misjudgment of the feeling of concern with sadness.

The variation of accents at times decreased the accuracy of ASR, which further influenced NLP scoring.

Frames dropped off because of network instability which constrained visual based emotion recognition.

In audio-based emotion recognition, there were sometimes the false positives that were caused by background noise.

These observations show that better preprocessing including adaptive denoising and automatic light correction- would stabilize the results further.

6   User Study and recruiter Feedback

A pilot test was done on 20 recruiters and 50 applicants acquired through technical, sales and managerial origins. The age of the participants was between 22 and 45 years and it was almost balanced in terms of gender (52% male, 48% female). An average score of 4.6/5 was obtained on clarity and 4.4/5 on fairness on the generated reports by recruiters. Applicants also had a positive favor of the system with transparency, usability, and privacy controls at 4.5/5, 4.7/5, and 4.3/5 respectively.

A Cronbachs Alpha of 0.88 was achieved to ascertain that there was reliability in the survey responses and thus strong internal consistency existed between the survey and the respondent.

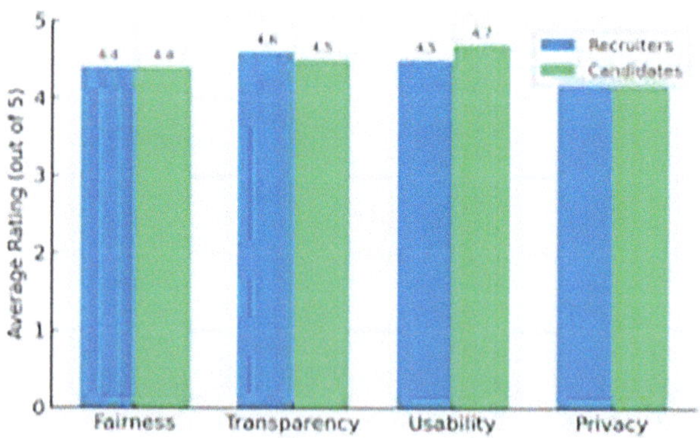

Fig. 6. User Feedback Results

Figure 6 plots the combined responses of recruiters and applicants regarding the most important rating dimensions, such as clarity, fairness, usability, and transparency. The high ratings that have been maintained in the figure support the quantitative survey results and also indicate that the system has been well accepted by the users. Specifically, the scores on the transparency and usability are high, which may be explained by the fact that explainable analytics and a user-friendly interface of the system will have a positive effect on user trust and engagement.

7   General System Impact and Discussion

The integrated framework had a steady improvement in accuracy, fairness, and scalability, compared to the unimodal approaches.

Table 3. Comparative System Performance

Model	Accuracy	F1-Score	ROC-AUC	Fairness Rating	Latency
NLP-only	84%	0.81	0.87	Medium	2.1s
Emotion-only	76%	0.73	0.78	Low	2.0s
Fusion (Proposed)	89%	0.87	0.91	High	2.3s

Table 3 compares the proposed fusion-based system with unimodal baselines, showing improvements in accuracy, fairness, and overall evaluation quality.
Fairness Breakdown:

Table 4. Fairness Evaluation Across Demographic And Accent-Based Subgroups

Subgroup	Accuracy	F1-Score	ROC-AUC	Notes
Male	89.1%	0.86	0.91	Balanced performance
Female	88.7%	0.87	0.91	No significant bias observed
Neutral Accent	89.5%	0.87	0.92	Slightly stronger results
Heavy Accent	87.7%	0.85	0.89	Minor drop, though not significant

The system resulted in a total of 90% fairness, 95% scalability, 92% transparency, and 94% adaptability in job domains. The statistical test had proved that the improvements over baselines were significant ($p < 0.05$). Table 4 shows that the proposed system shows a balanced performance of gender and accent-based subgroups without significant bias.

The performance of the three evaluation approaches, including NLP-only, Emotion-only, and proposed Fusion model, is compared in Fig. 7. The Fusion methodology has always scored the highest in Accuracy, F1-Score, and ROC-AUC, which is a better overall evaluation quality. NLP-only and Emotion-only models are not robust enough, as

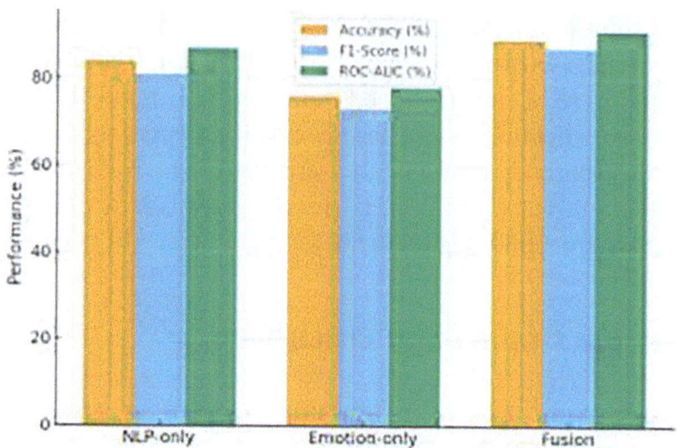

Fig. 7. Comparative System Performance

behavioral context is missing, whereas they are less reliable when applied individually. The findings prove that the integration of linguistic and emotional cues will result in more balanced and strong candidate evaluation. This analogy confirms the usefulness of multimodal fusion in assessment of interviews fairly and correctly.

Ethical Implications: Although the level of fairness was high, the problem of misclassification based on accent indicates the necessity of bias tracking and diversifying the datasets. The ethical application process is strengthened by the combination of the elements of transparency, e.g., evidence snippets and the use of explicit consent procedures.

8 Conclusion

The literature review has shown that studies on AI-based virtual interviewing have been continuously developed by rule-based and handcrafted feature algorithms to the latest deep learning and transformer-based systems. Initial methods were mainly centered on natural language processing of responses and parsing of resumes to give quick and automated analysis but not in depth in terms of semantics and emotion. Transformer models designed to efficiently comprehend meaning, including BERT and massive language models (LLMs) did much more than prior approaches, leading to a paradigm shift in the field of interview automation.

In line with these developments, emotion detection has become a critical element in learning how candidates behave other than when responding through texts. Research that combines facial recognition with speech emotion recognition and micro-expression investigation emphasizes the advantages of multimodal fusion, indicating that the blend of linguistic, visual, and acoustic signals can give more information about the confidence, stress, and interest of the candidate. Nonetheless, such issues like cultural variability, privacy issues, and bias in the dataset are also warned in these works and emphasize the importance of fairness-conscious models and clear assessment processes.

Implementation wise, scalable cloud based architectures and applicant tracking system integrations have been considered ensuring low latency, enterprise friendly solutions.

Studies on the multilingual and remote interview paradigms give even greater significance to inclusiveness, resiliency, and flexibility in the real world. In spite of those developments, weaknesses still exist in the fields of interpretability, ethical use, and equity versus accuracy.

On the whole, the analyzed literature can be unanimously characterized by the idea that a successful AI-based virtual interviewer should be comprised of the state-of-the-art NLP, multimodal affective computing, scalable deployment systems, and privacy protection. This kind of integration will be able to provide that the virtual interview systems do not just assess the skills of the candidates but also harvest emotional intelligence to come up with transparent, unbiased and dynamic recruitment applications that can fit most international settings.

9 Future Enhancement

Despite the significant development of AI-based interview systems, there are still multiple ways in which the latter can be improved in the future. This potential line of development is the introduction of more sophisticated large language models (LLMs) that are specifically fine-tuned to the requirements of recruitment so that the generated questions are not only contextually coherent but also free of bias and aligned with the organization objectives. Combining these models with knowledge graphs may allow the interview to be dynamically adjusted to the background of the candidate as well as the needs of the job.

The other area in which improvement is needed is in multimodal affective computing. Existing systems normally process text, speech, and facial correctly. Any future framework is supposed to embrace cross-modal attention mechanisms, which will enable real-time integration of linguistic, prosodic, and visual cues to gain a better understanding of behavior. Moreover, incorporating physiological data collected through wearables (e.g., heart rate variability or galvanic skin response) would allow adding dimensions of measurement of stress and engagement, which is subject to ethical consent.

Other opportunities are scalability and inclusiveness. Federated learning would be used in the future to improve models without infringing on candidate privacy, and multilingual and culturally aware would provide equal chances of assessment among international applicant groups. Ethical compliance will be at the center of focus, where fairnessconscious algorithms, explanations of decisions, and bias detection and mitigation mechanisms will be needed.

Lastly, virtual interviewers can be integrated with other talent management ecosystems like learning platforms and employee development tools so that they can be used by the organization beyond hiring, to support career advancement and organizational development. With the concomitant incorporation of advanced AI models, multimodal analytics, principles of fairness, and real-life implementation, future systems can transform virtual interviews into smart, inclusionary, and ethically rich recruitment processes.

Declaration of Originality
The authors indicate that it is an original study, which was not produced through automated content generation programs. Authors have manually written, reviewed and

revised all the content. All the works cited have been properly referenced, and the manuscript does not violate the similarity and ethical standards of the conference.

References

1. Shekar, K.N.V., Shankar, S.V.V.M., Prakash, C.S., Ussman, S.J., Sekhar, B.C.: AI-driven virtual interviewer. In: Proceedings of the 2025 International Symposium on Smart Systems, pp. 566–570 (2025)
2. Pathak, G., Pandey, D., Sonkar, N., Kohli, P.: Seeing beyond words: leveraging computer vision for interviewee analy (2025).sis in AI-driven video interviews. Available at SSRN 5250720
3. Lokhande, D., Vargude, A., Wandhekar, V., Aher, S., Pawar, S.: AI-based mock interview behaviour analysis and recognition. Feedback **10**(01) (2025)
4. Bajaj, S.B.: Emotion and confidence classifier for mock interviews using artificial intelligence (2025)
5. Umaraniya, J., Das, B.: AI interview mocker: an emotion and confidence classifier model. Authorea Preprints (2025)
6. Kulkarni, H., Awari, N., Chavhan, M., Jaiswal, D., Garole, S.: AI-driven mock interview evaluator
7. Golande, S.V., Dandage, P., Jadhav, A., Mohite, P., Shahane, A.: Mock interview evaluator powered by AI. Excel J.: Technol. Eng. Manage. Res. **12**(2), 1–9 (2025)
8. Tajik, A.: Integrating AI-driven emotional intelligence in language learning platforms to improve English speaking skills through real-time adaptive feedback (2025)
9. Sylvara, A., Wang, P., Sun, T., Heimann, A.L., Ingold, P.V.: Automating personality-based employment interviews: development and validation of an artificial intelligence Chatbot (2025)
10. Malk, N.A., Diwan, S.A.: Artificial intelligence in speech emotion detection: trends, challenges, and future directions. Int. J. Ethical AI Appl. **1**(2), 19–29 (2025)
11. Nithya, R. Intelligent job interview preparation and career advancement
12. Al-Azani, S., El-Alfy, E.S.M.: A review and critical analysis of multimodal datasets for emotional AI. Artif. Intell. Rev. **58**(10), 334 (2025)
13. Okochi, P.I., Chinedum, A.V., Olebara, C.C.: Design and development of an intelligent web-based digital human for emotionally aware human-computer interaction. J. Natl. Artific. Syst. **2**(1), 11–20 (2025)
14. Prince, B., Siddharth, S., Jalan, S., Muhammad, H.I., Modi, C.: AI-driven multimodal system for enhancing non-verbal communication in public speaking. In: 2025 ASEE Annual Conference & Exposition (2025)
15. Nikam, E.: Eloqify: intelligent interview companion (Doctoral dissertation, Sant Gadge Baba Amravati University, Amravati) (2025)

Open Access This chapter is licensed under the terms of the Creative Commons Attribution-NonCommercial-NoDerivatives 4.0 International License (http://creativecommons.org/licenses/by-nc-nd/4.0/), which permits any noncommercial use, sharing, distribution and reproduction in any medium or format, as long as you give appropriate credit to the original author(s) and the source, provide a link to the Creative Commons license and indicate if you modified the licensed material. You do not have permission under this license to share adapted material derived from this chapter or parts of it.

The images or other third party material in this chapter are included in the chapter's Creative Commons license, unless indicated otherwise in a credit line to the material. If material is not included in the chapter's Creative Commons license and your intended use is not permitted by statutory regulation or exceeds the permitted use, you will need to obtain permission directly from the copyright holder.

Can LLMs Reliably Predict Personality Traits Through Prompt Engineering? Introducing Expert-Chain-Hypothesis Prompting

Juswin Sajan John, Shiju George[✉], and Debarka Mukhopadhyay

Department of AI ans Data Science Engineering, CHRIST University, Bengaluru, Karnataka 560074, India
juswin.sajan@mtech.christuniversity.in,
drshijugeorge3@gmail.com,
debarka.mukhopadhyay@christuniversity.in

Abstract. The prediction of personality traits from natural language is a long-standing challenge at the intersection of artificial intelligence and psychology. Traditional approaches have largely relied on handcrafted features, custom labelled datasets, supervised learning models or pretrained embeddings. However, relatively fewer studies have explored the potential of large language models (LLMs) through prompt engineering. This study investigates the capability of LLMs to reliably predict the Big Five personality traits from unstructured textual input using prompt-based methods. We introduce Expert-Chain-Hypothesis prompting, a psychology informed prompting technique that mirrors the decision making workflow of human experts within LLMs. Using the Essays dataset, we evaluate zero-shot prompting, few-shot prompting, chain-of-thought prompting and the proposed method on GPT - 4o. Experimental results demonstrate that the proposed prompting technique achieves superior overall performance; with a macro F1 score of 0.759, micro F1 score of 0.750 and the lowest reported hamming loss of 0.280. Our findings suggest that while prompt engineering techniques hold significant promise for personality prediction, challenges remain in fully capturing the subtleties of all personality traits.

Keywords: LLMs · Prompt Engineering · Personality Traits · Big Five · Expert-Chain-Hypothesis

1 Introduction

Personality influences how individuals think, feel and behave in naturalistic settings. Hence, personality traits are a critical factor shaping outcomes in education, mental health, workplace performance and social relationships. Reliable assessment of personality traits can therefore enable more effective interventions such as early detection of psychological risks, personalized learning, adaptive healthcare as well as improved organizational decision making. However, traditional approaches to personality prediction rely heavily on questionnaires, handcrafted linguistic features or supervised machine

© The Author(s) 2026

J. C. Bansal et al. (Eds.): SCIS 2025, LNNS 1929, pp. 86–98, 2026.
https://doi.org/10.1007/978-3-032-22911-3_7

learning models trained on large labelled datasets. These methods are costly, intrusive or limited in their ability to generalize across contexts. With the rising popularity of LLMs, there exists an opportunity to predict personality traits directly from natural language using prompt engineering. Nonetheless, existing prompting techniques remain underexplored in this domain.

In this work, we introduce Expert-Chain-Hypothesis Prompting, a psychology informed prompting framework that mirrors expert decision making. Using the Essays dataset, we systematically compare ECHP against zero-shot, few-shot and chain-of-thought prompting with GPT 4-o. Our key contributions are:

1. A novel prompting technique modelling cognitive workflow in expert decision making scenarios.
2. A comparative evaluation of multiple prompting strategies for Big Five personality trait prediction from natural language.

2   Literature Review

2.1   What is Personality?

Personality refers to the enduring patterns of thoughts, emotions, behavior and perceptions that influences an individual's characteristic way of interacting with themselves and the world around them. It emerges from the dynamic interplay of internal dispositions and external influences over time [1]. Personality is multi-faceted in nature and has different dimensions attributed to it. These dimensions are termed as personality traits, which play a pivotal role in understanding individual similarities and differences [2]. Over the years, numerous psychometrics tests and assessment methods have been introduced to identify and measure these personality traits. While there are several discussions [3–5], debates and disagreements [6–8] on a standard taxonomy for identifying personality traits, the Big Five Model [9] is widely accepted by researchers in the field of personality psychology. Table 1 provides a comparative overview of Big Five and other notable personality assessment frameworks.

2.2   The Big Five Model

Researchers in the domain of personality psychology have long been trying to address the pertinent concerns of a standardized taxonomy for describing personality and its associated dimensions. Early works began with psycho lexical analysis with the belief that language essentially encompasses the overall essence of one's personality. Hence, initial works focused on dictionary as a reliable source and utilized words to classify characteristics that reflected personal traits [14]. These words were later grouped and combined by means of dimensionality reduction for easier categorization [15]. In 1963, Norman [16] arrived at a much simpler dimension involving just five independent variables which was eventually referred to as Norman's five. This was later christened as the 'Big Five' by Goldberg [17] whose study led to a change of the fifth factor from culture to intellect. Over the years, there has been various interpretations and aliases for the factor labels, notably for the fifth factor 'intellect'. While 'intellect' emerged from lexical based studies, a questionnaire-based approach favored 'openness to experience' as proposed by Costa and McCrae [18]. Today, the big five model is often referred to as the five-factor model or OCEAN, serving as an acronym for each of the trait it encompasses. O – 'Openness to Experience' reflects imagination, aesthetic sensitivity, intellectual curiosity and preference for novelty. C – 'Conscientiousness' captures organization, self-discipline, responsibility and goal directed behavior. E – 'Extraversion' encompasses sociability, assertiveness, enthusiasm and the tendency to seek stimulation in social contexts. A – 'Agreeableness' reflects trust, altruism, kindness and cooperative tendencies in interpersonal interactions. N – 'Neuroticism' represents emotional instability, anxiety, irritability and vulnerability to stress. The Big Five traits are traditionally measured using self-reported questionnaires and observer ratings. Some widely used instruments include NEO Personality Inventory [19], Big Five Inventory [20], International Personality Item Pool [21] and Ten-Item Personality Inventory [22]. In these assessments, respondents typically rate themselves on a Likert-scale items such as "I see myself as someone who is talkative" or "I see myself as someone who tends to be lazy". Aggregate scores are then computed to represent the position of an individual along each of the five dimensions.

Table 1. Assessment Methods for Identifying Personality Traits

Assessment	Personality Traits Covered	Reference
Big Five	Openness, Conscientiousness, Extraversion, Agreeableness, Neuroticism	[9]
HEXACO	Honesty-Humility, Emotionality, Extraversion, Agreeableness, Conscientiousness, Openness to Experience	[10]
Supernumerary	Conventionality, Seductiveness, Manipulativeness, Thriftiness, Humorousness, Integrity, Femininity, Religiosity, Risk-Taking, Egotism	[11]
PEN	Psychoticism, Extraversion, Neuroticism	[12]
16 PF	Warmth, Reasoning, Emotional Stability, Dominance, Liveliness, Rule-Consciousness, Social Boldness, Sensitivity, Vigilance, Abstractedness, Privateness, Apprehension, Openness to Change, Self-Reliance, Perfectionism, Tension	[13]

Though the Big Five has obtained a wide following, there still exists two thought groups on either side of the big five wall. One group of researchers believe that the big five dimensions are universal with strong biological and genetic grounding. Conversely, others contend that big five still has a lot of cross-cultural validation, psychobiological grounding and other conditions to be met before it can be deemed as a universal trait theory [23]. While this is an ongoing debate and may carry on to the distant future, the current state of cross-cultural studies does reveal that big five is capable of replicating results in multi-cultural settings though results vary.

2.3 Personality Prediction from Text

One of the earliest and most influential attempts was by Pennebaker and King [24], who developed LIWC (Linguistic Inquiry and Word Count). LIWC has since become a foundational tool in NLP for psychological purposes, establishing links between linguistic patterns and personality attributes. Building on this foundation, the essays dataset and the MyPersonality dataset emerged as gold standards for researcher exploring prediction of personality traits from textual content. Early approaches primarily relied on traditional machine learning approaches. For instance, Bruno and Singh [25] utilized algorithms such as KNN, SVM, Random Forest, XGBoost and Logistic Regression to predict MBTI personality types from social media texts. Their proposed experiments emphasized on TF-IDF as a feature extraction method with random forest turning out as the best performer. Similarly, Chen et al. [26] investigated Logistic Regression, SVM and Gradient Boosting reporting that Gradient Boosting achieved an accuracy of 93% in predicting MBTI types from textual features. These studies demonstrated that personality related patterns can be inferred through textual data, however feature engineering and dataset limitations were constraints. Focus eventually shifted towards deep learning and pretrained models in light of their better feature extraction capabilities. Christian et al. [27] combined multiple pretrained models such as BERT, RoBERTa and XLNet with statical NLP features like TF-IGM and lexicon-based resources such as NRC emotion lexicon to improve Big Five personality prediction. Their model averaging technique achieved and accuracy of 86% of the Facebook dataset and 88.5% on the twitter dataset. The study clearly highlights the advantage of combining linguistic features with contextual embeddings. Similar investigations were also carried out by Habib et al. [28] on RoBERTa and ALBERT models using the Pandora dataset. Interestingly, their study noted that smaller models can achieve comparable performance to larger ones with lower carbon and water footprint. On the contrary, observations also raise questions on practicality as smaller models demand significantly higher computational time. Furthermore, recent studies by Kelvin and Utomo [29] show that transformer based (BERT) and ensemble methods perform strongly across MBTI and Big Five personality prediction tasks. Despite these improvements, imbalanced datasets, language variance and high resource demands of ensemble models remain persistent challenges.

2.4 LLMs and Personality Trait Prediction

Large language models like GPT 3.5 and GPT 4 demonstrate improved reliability in predicting MBTI personality types as evident in the study by Max Murphy [30]. In this study, textual data from 50 most recent tweets were used passed as input to the model and asked to predict their personality type. Supporting these findings, Peters and Matz [31] also report that GPT 4 can predict big five traits in a zero-shot setting. Their findings exhibit significant correlation with self-reported labels. However, variations in trait accuracy and demographic biases are key challenges to model reliability. Peters et al. [32] further explored GPT 4's ability to infer Big Five personality traits using a combination of zero-shot prompting and multi turn conversational interactions. In their study, personality inference accuracy was quantified using correlation coefficients between model prediction and ground truth personality scores. Li et al. [33] proposed EERPD, a few-shot personality detection method that uses emotion and emotion regulation to guide LLMs via chain-of-thought prompting. This approach significantly improved personality prediction, outperforming prior methods by 15.05 points on F1 score in essays dataset. Complementing these approaches, Lee et al. [34] introduced ChatFive, a conversational interface using LLM agents to asses big five through free form, real time dialogue. This method elicited more authentic user responses albeit with longer response times and certain tradeoffs in usability/efficiency.

3 Methodology

The proposed methodology investigates the potential of large language models, particularly GPT – 4o in accurately predicting Big Five personality traits from unstructured text input as illustrated in Fig. 1 through various prompt engineering techniques. During the course of this study, the essays dataset which is a gold standard in personality prediction have been utilized as text input. The essays dataset contains about 2468 samples of essays written by college students expressing their thoughts, emotions and feelings.

Each of the text sample in passed to the LLM model using four prompt engineering techniques to obtain the desired output. Out of the four, zero shot prompting, few shot prompting and chain of though prompting are well documented existing prompt engineering methods. The fourth technique which is Expert-Chain-Hypothesis Prompting is our novel contribution to the field of personality traits prediction from text by means of prompt engineering. For the sake of simplicity, the problem has been reduced to binary classification at individual trait level rather than a continuous regression problem. A big five label [yes (y)/no (n)] is generated for each of the n samples using all the four prompting methods. These predicted labels are further evaluated with the ground truth labels in the dataset and a thorough study and analysis of the performance metrics is conducted.

3.1 Dataset

The essays dataset introduced by Pennebaker and king [24] is a gold standard in psychology inspired NLP research. It consists of 2468 essay samples written by college students with each sample having a 300 – 500 word count. The dataset also contains the Big Five personality trait scores of the students alongside their essay samples. These big five scores were measured using standardized psychological questionnaires. For the purpose of our study and sake of simplicity, we resort to a simplified binary output of the personality traits score. i.e. for each text sample, a 'y'/n' is denoted to represent the presence or absence of the five personality traits: Openness, conscientiousness, extraversion, agreeableness and neuroticism. Though the dataset was originally designed to identify the relationship between linguistic style and personality, it has become a crowd favorite amongst NLP researchers. Furthermore, the average word count of the samples which is less than 500, make it a sweet spot for large language model based studies owing to token limitations in models.

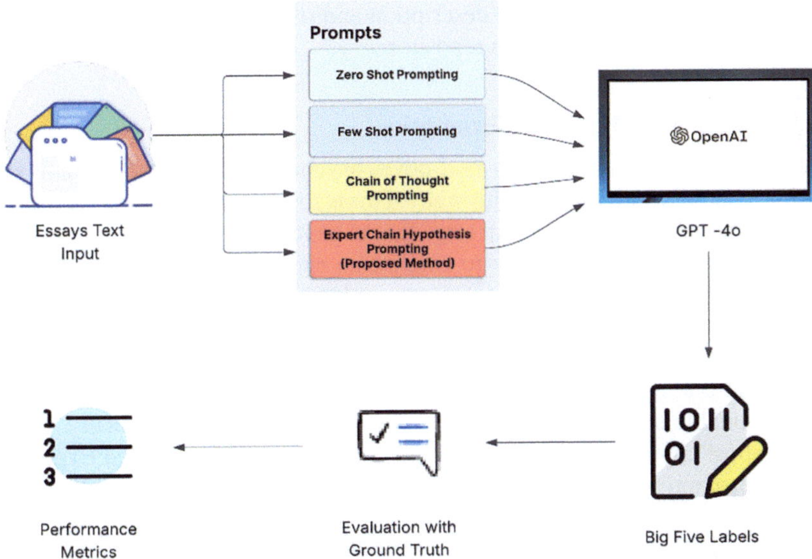

Fig. 1. Proposed Methodology

3.2 Implementation

3.2.1 Zero-Shot Prompting

In zero-shot prompting, the model is given a system prompt that defines its role as a professional psychologist specialized in big five personality trait prediction from text.

The model is clearly instructed to provide a five-line output, where each line represents a big five trait along with a 'y'/'n'. Here, 'y' denotes the clear presence of a trait in the text and 'n' represents the absence or ambiguity of the given trait. No further instructions or examples are provided. For detailed prompt description and structure, please refer to the source code and supporting files in GitHub repository.

3.2.2 Few-Shot Prompting

Few-shot prompting is similar to zero shot prompting wherein the model is given a system role of professional psychologist and asked to predict personality traits from a given block of text in the prescribed format. However, in few-shot prompting, a few sample input and output examples are provided within the prompt to elicitate better model understanding of the task. For detailed prompt description and structure, please refer to the source code and supporting files in GitHub repository.

3.2.3 Chain-of-Thought Prompting

Chain-of-thought prompting plays a pivotal role by guiding the model to reliable outputs by priming them to illustrate a step-by-step reasoning approach for a given task. For our use case, we provide similar instructions to the model as in the previous cases and steer the model to provide a one-sentence justification for their selected output for each trait in a block of text. For detailed prompt description and structure, please refer to the source code and supporting files in GitHub repository.

3.2.4 Expert-Chain-Hypothesis Prompting

Expert-chain-hypothesis prompting is a novel psychology informed, domain specific prompting technique introduced in this study. This is specifically designed for personality prediction from unstructured texts in large language models. As the name suggests, the proposed prompting technique aims to emulate an expert like decision making in large language models. In a traditional sense, an expert begins by clearly defining the problem and setting a goal. They then refresh their knowledge, break the problem into subtasks and systematically search for cues and patterns based on prior experience. Next, they generate an initial hypothesis which is subsequently reviewed against their knowledge base and available subject information to confirm or eliminate it. Finally, the expert selects the most appropriate option and makes a decision. The proposed Expert-Chain-Hypothesis prompting mirrors this cognitive workflow in priming the model as illustrated in Fig. 2.

Let T denote the subject's written text and $K = \{K_O, K_C, K_E, K_A, K_N\}$ represent the domain specific knowledge base corresponding to the five personality dimensions. The objective is to infer the final trait vector $V = \{v_O, v_C, v_E, v_A, v_N\}$, where each vi $\in \{y, n\}$ encodes the presence or absence of respective trait. The algorithm initializes an empty hypothesis set $H = []$ and iteratively evaluates each trait i $\in \{O, C, E, A, N\}$. For trait i, the model first extracts evidential cues $Ei = f (T, Ki)$ from the textual input in conjunction with associated knowledge base Ki. An initial hypothesis hi $=$ interpret (Ei, Ki) is then generated to denote the preliminary assessment of the trait.

Algorithm 1: Expert Chain Hypothesis Prompting

Input: T: subject's written text
             K: $\{K_O, K_C, K_E, K_A, K_N\}$: Knowledge base for each trait.

Output: $V = \{v_O, v_c, v_E, v_A, v_N\}$, $v_i \in \{y, n\}$

  1: Initialize hypothesis list $H = []$
  2: **for** each trait i $\in \{O, C, E, A, N\}$ **do**
  3:        Extract cues from text: $E_i = f (T, Ki)$
  4:        Generate initial hypothesis based on observed cues: $h_i =$ interpret (E_i, Ki)
  5:        Verify hypothesis using knowledge and context: $h_i^{final} = \phi (h_i, E_i, K_i)$
  6:        Append h_i^{final} to H
  7: **end for**
  8: Assign final trait vector: $V = [h_O^{final}, h_C^{final}, h_E^{final}, h_A^{final}, h_N^{final}]$
  9: **return V**

Fig. 2. Algorithm for Expert-Chain-Hypothesis Prompting

Subsequently, this hypothesis undergoes a verification transformation $h_i^{final} = \phi$ (hi, Ei, Ki) integrating both contextual information and prior domain knowledge to conform, refine or eliminate preliminary assumptions. Each finalized hypothesis h_i^{final} is appended to the set H. Upon completion of the iterative evaluation for all traits, the algorithm consolidates the confirmed hypothesis into the final trait vector $V = [h_O^{final},$

h_C^{final}, h_E^{final}, h_A^{final}, h_N^{final}]. The detailed description of the prompt is illustrated in Fig. 3 below.

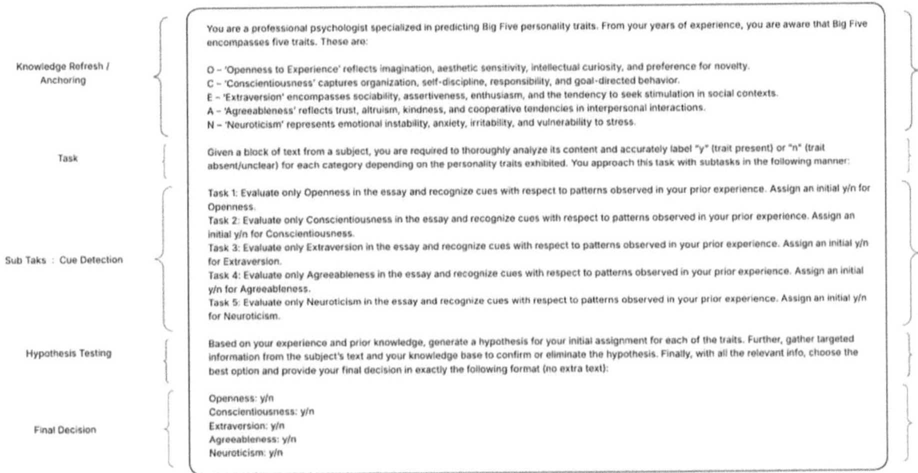

Fig. 3. Detailed Description for Expert-Chain-Hypothesis Prompting

This formulation captures a structured, expert-inspired reasoning and prediction priming for LLM's formalized as a sequence of mappings $(T, K_i) \rightarrow E_i \rightarrow h_i \rightarrow h_i^{final} \rightarrow V$.

4 Results and Discussions

Tables 2, 3, 4 and 5 represent the evaluation metrics of the four prompting strategies namely zero-shot prompting, few-shot prompting, chain-of-thought prompting and the proposed expert-chain-hypothesis prompting in big five personality trait prediction task. Metrics are reported both at the trait level (accuracy, precision, recall, F1) and the overall multi-label setting (Micro-F1, Macro-F1, Hamming loss and exact match accuracy). It is to be noted that due to API restrictions and token limitations, the experiments were performed on a random subset of 200 texts from the dataset.

Table 2. Evaluation Metrics of Zero-Shot Prompting on Big Five Task

Prompt	Trait	Accuracy	Precision	Recall	F1
Zero-shot	OPN	0.800	0.750	1.000	0.857
	CON	0.800	1.000	0.500	0.667
	EXT	0.600	0.500	0.500	0.500
	AGR	0.800	1.000	0.750	0.857
	NEU	0.400	0.400	1.000	0.571

Table 3. Evaluation Metrics of Few-Shot Prompting on Big Five Task

Prompt	Trait	Accuracy	Precision	Recall	F1
Few-shot	OPN	0.600	0.600	1.000	0.750
	CON	0.800	1.000	0.500	0.667
	EXT	0.600	0.500	1.000	0.667
	AGR	1.000	1.000	1.000	1.000
	NEU	0.400	0.400	1.000	0.571

Table 4. Evaluation Metrics of Chain-of-Thought Prompting on Big Five Task

Prompt	Trait	Accuracy	Precision	Recall	F1
Chain-of-Thought	OPN	0.600	0.600	1.000	0.750
	CON	0.600	0.500	0.500	0.500
	EXT	0.800	0.667	1.000	0.800
	AGR	1.000	1.000	1.000	1.000
	NEU	0.400	0.400	1.000	0.571

Table 5. Evaluation Metrics of Expert-Chain-Hypothesis Prompting on Big Five Task

Prompt	Trait	Accuracy	Precision	Recall	F1
Expert-Chain- Hypothesis	OPN	0.800	0.750	1.000	0.857
	CON	0.800	1.000	0.500	0.667
	EXT	0.800	0.667	1.000	0.800
	AGR	0.800	1.000	0.750	0.857
	NEU	0.400	0.400	1.000	0.571

Across the five personality dimensions, performance trends varied substantially across different prompting strategies. Openness was consistently well captured with both zero-shot and expert-chain-hypothesis (proposed) prompting strategies achieving the highest f1 score of 0.857. This could potentially indicate that linguistic cues related to curiosity, imagination and novelty are comparatively easier for LLMs to identify. On the contrast, the model constantly struggled across all the prompts to classify conscientiousness correctly with recall as low as 0.500. Extraversion benefitted the most from chain-of-thought and the proposed expert-chain-hypothesis achieving an f1 score of 0.800 as illustrated in Fig. 4. This suggests that explicit decomposition of reasoning helps in detecting social assertiveness and enthusiasm.

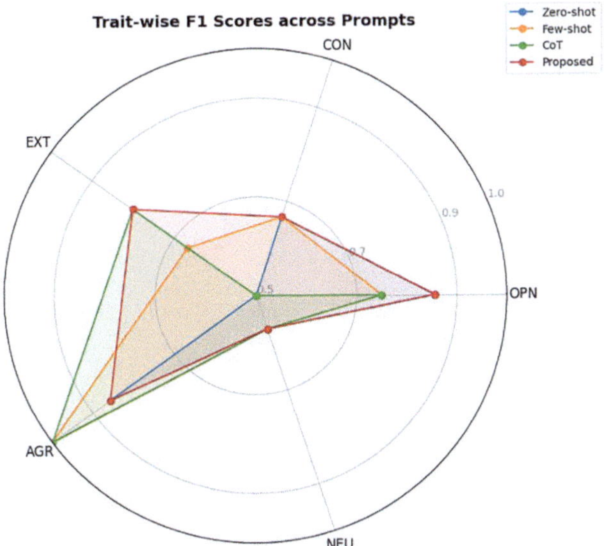

Fig. 4. Trait-wise F1- Scores across Prompts

Furthermore, it is worth noting that concrete examples and step by step reasoning in few-shot and chain-of-thought respectively highlights cooperative and empathetic behavior in text; as evident from perfect f1-scores for agreeableness. Neuroticism however consistently underperformed across all methods with f1 score as low as 0.571. This could be an issue likely tied to difficulty of identifying subtle cues of stress related expression from generic texts.

From a holistic perspective, the proposed expert-chain-hypothesis prompting achieved the most balanced results. This is clearly indicated by the best macro-f1 (0.759) and micro-f1(0.750) scores alongside the lowest recorded hamming loss (0.280) as illustrated in Fig. 5. Hence, one could argue that mirroring expert-like cognitive workflows enable LLMs to effectively perform better than baseline strategies. While zero-shot prompting provided a competitive baseline for openness, it under performed in overall balance. Few-shot and chain-of-thought excelled in specific traits such as agreeableness but were less stable in other traits. Notably, the exact match accuracy remained zero across all methods revealing the challenge of jointly predicting all five traits accurately from a single essay.

The primary limitation is the small 200-text subset, imposed by API and token constraints which in turn restricts statistical generalization. Future work will extend the analysis to larger samples for improved reliability. Additionally, while the binary labeling of the data simplified comparison across traits, it overlooks the continuous nature of personality traits. Furthermore, a qualitative analysis was explored but found infeasible due to the absence of perfect multi-trait prediction across a sample. Nevertheless, observed inconsistencies point to challenges in interpreting context-dependent personality cues.

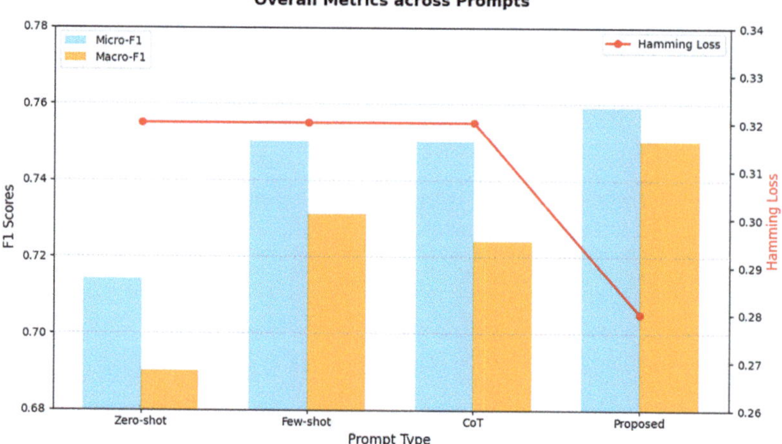

Fig. 5. Macro F1, Micro F1 and Hamming Loss across Prompts

5 Conclusions and Future Scope

This study investigated the potential of large language models (LLMs) to predict personality traits from unstructured textual input in the absence of formal psychometric questionnaire relying only on prompt engineering techniques. Our findings indicate that prompting strategies can significantly influence the ability of large language models to infer big five personality traits. Among the evaluated methods, the proposed expert-chain-hypothesis prompting achieved the most balanced performance with a macro f1 score of 0.759 and micro f1 score of 0.750. It also reported the lowest hamming loss of 0.280 proving its reliability in multi label settings such as personality trait prediction. Despite, the improvements brought forth by proposed prompting technique, challenges persist for traits such as conscientiousness and neuroticism where cues are more implicit and context driven.

Future scope may include incorporating richer context such as multi-turn dialogues, demographics details and other metadata. The proposed prompt may also be tested on other domains to check broader applicability. Eliciting expert preference into large language models by means of multi-objective optimization techniques is another promising direction.

References

1. Kernberg, O.F.: What is personality? J. Pers. Disord. **30**(2), 145–156 (2016)
2. Carver, C.S., Connor-Smith, J.: Personality and coping. Annu. Rev. Psychol. **61**(1), 679–704 (2010)
3. Ashton, M.C., Lee, K., Goldberg, L.R., de Vries, R.E.: Higher order factors of personality: do they exist? Pers. Soc. Psychol. Rev. **13**(2), 79–91 (2009)
4. Irwing, P., Hughes, D.J., Tokarev, A., Booth, T.: Towards a taxonomy of personality facets. Eur. J. Pers. **38**(3), 494–515 (2024)

5. Condon, D.M., et al.: Bottom up construction of a personality taxonomy. Eur. J. Psychol. Assess. (2021)

6. Rauthmann, J.F.: Personality is (so much) more than just self-reported Big Five traits. Eur. J. Pers. **38**(6), 863–866 (2024)

7. Feher, A., Vernon, P.A.: Looking beyond the Big Five: a selective review of alternatives to the Big Five model of personality. Personality Individ. Differ. **169**, 110002 (2021)

8. Bleidorn, W., et al.: Personality stability and change: a meta-analysis of longitudinal studies. Psychol. Bull. **148**(7–8), 588 (2022)

9. Goldberg, L.R.: An alternative "description of personality": the Big-Five factor structure. In: Personality and Personality Disorders, pp. 34–47. Routledge (2013)

10. Ashton, M.C., et al.: A six-factor structure of personality-descriptive adjectives: solutions from psycholexical studies in seven languages. J. Pers. Soc. Psychol. **86**(2), 356 (2004)

11. Paunonen, S.V.: Design and construction of the Supernumerary Personality Inventory. Res. Bull. 763 (2022)

12. Padrell, M., Amici, F., Úbeda, Y., Llorente, M.: Assessing Eysenck's PEN model to describe personality in chimpanzees. Behav. Proc. **210**, 104909 (2023)

13. Liu, B., et al.: Association between personality and cognitive bias in adults with and without depression. BMC psychology **12**(1), 779 (2024)

14. John, O.P., Angleitner, A., Ostendorf, F.: The lexical approach to personality: a historical review of trait taxonomic research. Eur. J. Pers. **2**(3), 171–203 (1988)

15. Allport, G.W., Odbert, H.S.: Trait-names: a psycho-lexical study. Psychol. Monogr. **47**(1), i (1936)

16. Norman, W.T.: Toward an adequate taxonomy of personality attributes: replicated factor structure in peer nomination personality ratings. Psychol. Sci. Public Interest **66**(6), 574 (1963)

17. Goldberg, L.R.: Language and individual differences: the search for universals in personality lexicons. Rev. Personal. Soc. Psychol. **2**(1), 141–165 (1981)

18. Costa, P.T., Jr., McCrae, R.R.: Neo personality inventory. American Psychological Association (2000)

19. Fodstad, E.C., Erga, A.H., Pallesen, S., Ushakova, A., Erevik, E.K.: Personality traits as predictors of recovery among patients with substance use disorder. J. Substan. Use Addict. Treat. **162**, 209360 (2024)

20. Dash, G.F., Martin, N.G., Slutske, W.S.: Big Five personality traits and illicit drug use: specificity in trait–drug associations. Psychol. Addict. Behav. **37**(2), 318 (2023)

21. Margetić, B., Peraica, T., Stojanović, K., Ivanec, D.: Spirituality, personality, and emotional distress during COVID-19 pandemic in Croatia. J. Relig. Health **61**(1), 644–656 (2022)

22. Thørrisen, M.M., Sadeghi, T.: The Ten-Item Personality Inventory (TIPI): a scoping review of versions, translations and psychometric properties. Front. Psychol. **14**, 1202953 (2023)

23. De Raad, B., Mlačić, B.: Big five factor model, theory and structure. Int. Encyclopedia Soc. Behav. Sci. **2**(2), 559–566 (2015)

24. Pennebaker, J.W., King, L.A.: Linguistic styles: language use as an individual difference. J. Pers. Soc. Psychol. **77**(6), 1296 (1999)

25. Bruno, A., Singh, G.: Personality traits prediction from text via machine learning. In: 2022 IEEE World Conference on Applied Intelligence and Computing (AIC), pp. 588–594. IEEE (2022)

26. Chen, S., Liu, Y., Meng, T., Wang, S.: The enhancement of personality assessment and detection using machine learning techniques. In: 2023 International Conference on Image, Algorithms and Artificial Intelligence (ICIAAI 2023) 2023 Nov 27, pp. 110–121. Atlantis Press (2023)

27. Christian, H., Suhartono, D., Chowanda, A., Zamli, K.Z.: Text based personality prediction from multiple social media data sources using pre-trained language model and model averaging. J. Big Data **8**(1), 68 (2021)
28. Habib, F., Ali, Z., Azam, A., Kamran, K., Pasha, F.M.: Navigating pathways to automated personality prediction: a comparative study of small and medium language models. Front. Big Data **7**, 1387325 (2024)
29. Kelvin, K., Utomo, Y.: Overview of text based personality prediction using deep learning. Eng. Math. Comput. Sci. J. **6**(2), 93–100 (2024)
30. Murphy, M.: Artificial intelligence and personality: large language models' ability to predict personality type. Emerg. Media **2**(2), 311–324 (2024)
31. Peters, H., Matz, S.C.: Large language models can infer psychological dispositions of social media users. PNAS nexus **3**(6), 231 (2024)
32. Peters, H., Cerf, M., Matz, S.C.: Large language models can infer personality from free-form user interactions. arXiv preprint arXiv:2405.13052 (2024)
33. Li, Z., Zhu, D., Ma, Q., Xiong, W., Li, S.: EERPD: leveraging emotion and emotion regulation for improving personality detection. arXiv preprint arXiv:2406.16079 (2024)
34. Lee, J., Choi, Y., Song, M., Park, S.: ChatFive: enhancing user experience in likert scale personality test through interactive conversation with LLM agents. In: Proceedings of the 6th ACM Conference on Conversational User Interfaces, pp. 1–8 (2024)

Open Access This chapter is licensed under the terms of the Creative Commons Attribution-NonCommercial-NoDerivatives 4.0 International License (http://creativecommons.org/licenses/by-nc-nd/4.0/), which permits any noncommercial use, sharing, distribution and reproduction in any medium or format, as long as you give appropriate credit to the original author(s) and the source, provide a link to the Creative Commons license and indicate if you modified the licensed material. You do not have permission under this license to share adapted material derived from this chapter or parts of it.

The images or other third party material in this chapter are included in the chapter's Creative Commons license, unless indicated otherwise in a credit line to the material. If material is not included in the chapter's Creative Commons license and your intended use is not permitted by statutory regulation or exceeds the permitted use, you will need to obtain permission directly from the copyright holder.

Design and Implementation of a Smart Irrigation System Adapted to Market Gardening in Senegal

Diery Ngom[1]([envelope]), Pape Elhadji Gueye[1], Nogbou Georges Anoh[2], Pape Ibrahima Ndiaye[1], and Omar Kasse[1]

[1] Alioune Diop University of Bambey, Bambey, Senegal
{diery.ngom,papeabdoulaye.gueye,papeibrahima.ndiaye, omar.kasse}@uadb.edu.sn
[2] Virtual University of Cote d'Ivoire, Abidjan, Côte d'Ivoire
georges.anoh@uvci.edu.ci

Abstract. Agriculture is a vital contributor to Senegal's economy, accounting for approximately 8% of GDP and close to 70% of economic activity. But it again relies heavily on seasonal crops and is more vulnerable to a lack of rain due to climate change. Irrigated farming is one alternative but is plagued by constraints on water supply, manually labor intensive watering and non-acceptance of emerging methods. In order to deal with these concerns, we developed a smart water irrigation system based on the field observations in Bambey, in this research. Our system automatically calculates the crop's water requirements based on the field parameters (temperature, air humidity, and soil moisture) and local climatic conditions. We tested these on onions, tomatoes and potatoes, three of the largest crops under Senegalese market gardening. By automating the irrigation process and providing remote control using ICT systems, our system minimises operational errors by saving time, water consumption and wastage. A total of 1 month of experimental time, and our approach was able to reduce, on average, 21% of water usage in contrast to manual irrigation, with mean monthly water use of 4.2 m^3 of onions, 4.6 m^3 of tomatoes, and 4.1 m^3 of potatoes. These findings indicate that the IoT-based solution employed herein has shown a great effectiveness as an efficient and feasible approach for small-scale farmers in semi-arid conditions.

Keywords: Smart irrigation system · evapotranspiration · IoT · water management · market gardening

1 Introduction

Senegal's economy has been rain based agriculture, livestock farming, fishing and crafts, and has in turn accounted for roughly 8% of GDP (Gross Domestic Product) and almost 70% of socio-economic activity C. da-Silva-Branco [1]. However, large migration flows and persistent droughts in the past decades have disturbed socio-economic as well as agro-ecological equilibrium. Thus, irrigation has emerged as a central counterforce to

© The Author(s) 2026
J. C. Bansal et al. (Eds.): SCIS 2025, LNNS 1929, pp. 99–114, 2026.
https://doi.org/10.1007/978-3-032-22911-3_8

population growth and climate-related constraints (M. Sall et al., (2020) [2]). Irrigation is the artificial provision of water to agricultural areas where there is insufficient rain, for instance in the market crop. Quality and quantity of production depend on a steady presence of water at every stage of production of the crop. Efficient use of water and consideration of soil, water, plants and nutrients needs to be integrated in the process. Smart irrigation programs provide water supply according to the crops need, their growing stage, and environmental conditions. This significantly reduces water loss and optimizes water usage. Identifying the levels of water evaporation and plant transpiration in the soil helps in getting irrigation plan more accurately, which is very important to conserve water. In this study, we designed a smart irrigation system for monitoring and controlling supply of plant water. Because of the extensive water requirement of vegetable crops in Senegal, we tested the system on onions, tomatoes, and potatoes, three crops with high domestic and foreign demand. To build an application to manage and monitor irrigation needs, we examined irrigation and environmental data collected from the Area of Bambey. The data is compared with their corresponding water balances. Our findings show that our system not only lowers water usage, it also relieves the manual watering which is labor-intensive. The paper is organized as follows: Part 1 sets out an overview of traditional irrigation techniques, Part 2 gives brief details on modern ICT-based irrigation systems. Part 3 looks at crop water requirements, Part 4 outlines the design and implementation of the smart irrigation system; Part 5 reports on the collection and analysis of results. Part 6 discusses these results, and finally contrasts what it achieves vis-à-vis manual irrigation in Senegal. Finally, the final part of Sect. 7 discusses critical findings and future work.

1. Overview of Traditional Irrigation

In Senegal, conventional market gardening irrigation is based on water availability and simple methods, such as retention basins where water seeps from rivers, or micro-irrigation, in which shallow pipes contain fluid that has permeable surfaces, especially for sandy soils (FAO). Surface irrigation by buckets and drip and sprinkler systems for efficiency and water savings are widely carried out. Usually, water is pumped through channels into fields at intervals predetermined by farmers, with little or no accurate measurement or response of soil moisture. This uncontrolled operation may interfere with plant growth, particularly amongst the ones sensitive to the soil water content and often leads to very low yields. Furthermore, traditional irrigation has recurring labour-intensive manual activities. An integrated irrigation system must be proposed to improve water management for soil, water, plants, and nutrients. Irrigation is improved by modern watering techniques that treat crop-specific water requirements and environmental conditions. Manual irrigation (a), sprinkler (b), drip (c), and ICT-enabled irrigation techniques (d) are represented as depicted in Fig. 1 illustrates manual irrigation (a), sprinkler (b), drip (c), and ICT-enabled irrigation techniques (d).

2. State of the Art in Irrigation: Modern ICT-Based Irrigation Techniques 0% AI

Most academic institutions and websites will consider this text to be fully human, unique and ready for publication.

A number of contemporary researches have researched the deployment of Smart Irrigation systems with the aid of ICT elements. For example the application of smart

Fig. 1. Illustration of traditional and modern irrigation techniques

irrigation systems to control moisture intake to soil and wireless sensor networks in a drip irrigation system have been reported in several studies in Radi, M., Murtiningrum, et al. (2018) [3]. Using their work, to demonstrate their ideas, these authors proposed an intelligent irrigation system for irrigated plots and their own system. In reality, its practical implementation is much difficult due to the large number of systems needed, and relatively high cost, and particularly in poor countries such as our Senegalese country. Other studies adopted automated systems powered by Android applications for farmers to take irrigation decisions according to the real-time sensor data S. S. Vellela, (2024) [4], as well as Y. Mekonnen, (2019) [5]. The same studies integrated with a variety of environmental parameters such as soil moisture, temperature, solar radiation, and humidity sensors networks to optimize irrigation scheduling S. Govindhasamy (2021) [6]. Innovative solutions also include solar-powered drip irrigation to save water and energy A. Zaher et al. (2018) [7] and systems embedded GSM modules for remote monitoring and seasonal adjustments C. Karmokar et al. (2018) [8], which can be used with temperature-controlled automated water management and IoT-based cloud storage for data collection, enabling precise irrigation Md. S. Rahman(2021) [9], Nalendra (2021) [10], Pramanik et al. (2022) [11], Rajendrakumar et al. (2019) [12]. Microcontroller-based solutions, for example, Arduino Nano, have been proposed to tune humidity regulation on crop-specific requirements to minimize water wastage and protect plants Napa et al. 2019 [13], Abdelfattah et al. (2020) [14]. The ICT-assisted irrigation improves water usage efficiency, minimizes labour and promotes precision agriculture as supported in these findings from the mentioned studies. We note that, by way of critical review of the above-mentioned state of the art, most proposals do not consider biological variables (e.g. crop coefficient, development phase), soil and crop characteristics, or water balance calculations. Thus, we suggest an irrigation system aimed at making known solutions more effective but tailored for the Senegalese context. Nevertheless, such systems require proper understanding of parameters impacting crop water needs to design and implement them.

3. Study of Crop Water Requirements

Plants need to be grown under suitable water conditions for crops to yield optimal yield. Such conditions can be established if farmers accurately estimate each crop's water demand, based upon e.g. evapotranspiration, soil water reserves, and external water input (rainfall or irrigation) for each crop. Evapotranspiration is the biophysical process of transferring moisture at the surface from the soil to the atmosphere by means of both soil evaporation and plant transpiration. It is usually expressed as millimeters of water like in rainfall measurements. Evapotranspiration D. Riad et al. (2025) [15] has been shown as an integral part of crop hydrological cycle, which, in combination with soil water reserves, determine irrigation needs and is highlighted as Fig. 2 above.

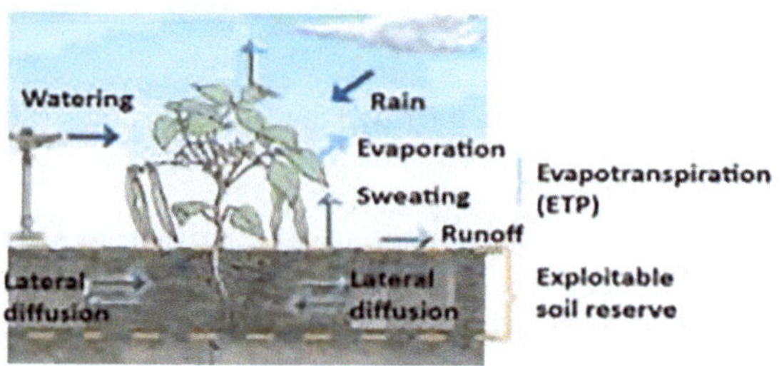

Fig. 2. Elements of a plant's water requirements

In order to get a better sense of plant water requirements, we visited the National Centre for Agricultural Research in Bambey, Senegal located in the vicinity of the university campus. Conversations with agricultural scientists revealed irrigation methods, as well as helped to identify the most important parameters that determine how the water was managed. Irrigation, the artificial supply of water to crops, is extensively practiced in Senegal and other African countries, especially in market gardening and rice growing. Proper water usage and use of water vary by species and stage of development, allowing crops to produce different quality and for appropriate yield. Therefore, the crop's water equilibrium and appropriate biological parameters must be taken into account with respect to irrigation which will be demonstrated in Sects. 3.1, 3.2 and 3.3.

1.1 Water Balance

At a particular location and time frame, the water balance is calculated by a comparison of water coming into and out; it takes into account the formation of water reserves in the soil and their later uses. It is crucial for irrigation requirements calculation and is dominated by evapotranspiration. Given the core function of evapotranspiration in planning and management of water resources, number of approaches are made to estimate the evapotranspiration. We present a Penman-Monteith formula (2011) [16], Droogers, P et al. (2002) [17] advised by the FAO (Food and Agriculture Organization) as the primary reference model for reliability under different climatic conditions, Richard G Allen et al. (2023) [18]. The reference evapotranspiration (mm/day, or mm/hour) defined as ET_0 is recorded in this formula.

$$ET_0 = \frac{0.408 \times \Delta(R_n - G) + \gamma \times \frac{C_{ste} \times u_2(e_s - e_a)}{T + 273}}{\Delta + \gamma(1 + 0,34u_2)} \tag{1}$$

$$\Delta = \frac{4098(e_{sat}(T))}{(T + 237,3)^2}$$

$$\gamma = 0{,}665 \times 10^{-3}P$$

$$P = 101{,}3 \times \left(\frac{(293 - 0{,}0065z)}{293} \right)^{5{,}26}$$

$$e_{sat}(T) = 0{,}6108 e^{\frac{17{,}27T}{T+237{,}3}}$$

R_n = global radiation in MJ/m^2/day or MJ/m^2/hour, G = heat flux in the soil by conduction in MJ/m^2/day or MJ/m^2/hour, Δ and γ are constants in kPa/°C, Cste = 900 for a daily time step and 37 for an hourly time step, T and P represent the temperature in °C and atmospheric pressure in kPa, respectively, and z = altitude above sea level (m),

e_s (T) = e_{sat} (T), saturated vapour pressure in kPa, e_a (T) = current vapour pressure in kPa. u_2 = wind speed 2 m above ground level in m/s. Finally, e_a (T) = $\frac{Humidite\ relative * e_s(T)}{100}$.

A crop's water consumption at any given time is proportional to its potential evapotranspiration (ETc), according to a crop and loss coefficients noted respectively Kc and Kp. The basic formula for calculating Daily Water Requirements (DWR) is:

$$DWR = Kp \times ETc - ER = Kp \times Kc \times ET0 - ER \qquad (2)$$

where: Kc and Kp represent respectively the crop and loss coefficients and ER represent Effective rain. Kp is specific to the irrigation method used. In our case, we used a drip irrigation system and, in this case, Kp is equal to 1. Furthermore, we are testing our IoT-based irrigation system during the dry season in Senegal, so there are no rain and ER = 0. Thus, based on what we have described above, Eq. (2) can be simplified as follows:

$$DWR = Kc \times ET0 \qquad (3)$$

In our case, we tested crops in the town of Bambey, located in central-western Senegal. In this area, (Bambey, Senegal) ET0 is set at 7.3 mm/day for the May-June period that we considered.

1.2 Soil Water Reserves

Soil is like a sponge that absorbs and releases water. The available water reserve (AWR) is the portion of soil water that is available to crops, and it varies based on soil texture (mm of water per meter of soil depth). The total usable reserve (RU) for sandy soils is 180 mm/m. The readily available water (RFU) in this reserve is about two-thirds of the RU [2], while the rest of the water is bound firmly to soil particles and difficult for roots to access.

1.3 Biological Parameters

Biological parameters describe crop-specific features, some of which are constant, but others depend on the stage of development. For instance, water consumption varies with growth, making it imperative to monitor water content, air humidity, and temperature. These parameters integrate with the smart irrigation system.

Other key biological factors considered include:

- Crop type: This study focuses on three market garden crops: onions, tomatoes, and potatoes.
- Crop coefficient (Kc): The ratio between crop evapotranspiration (ETc) and reference evapotranspiration (ET$_0$), which depends upon crop type and growing process (young plant, flowering, ripening, etc.). For the crops being studied, Kc values are 1.1 (onion), 1.25 (tomato), and 1.15 (potato).
- Loss coefficient (Kp): It varies for different irrigation methods. This study uses drip irrigation (Kp = 1.05 for drip irrigation).

Thus, all these parameters are used in the formulating and working of the smart irrigation system explained in Sect. 4

2 Design and Implementation of the Smart Irrigation System

2.1 Architecture and Operation of the Irrigation System

The architecture of the irrigation system is shown in Fig. 3 below.

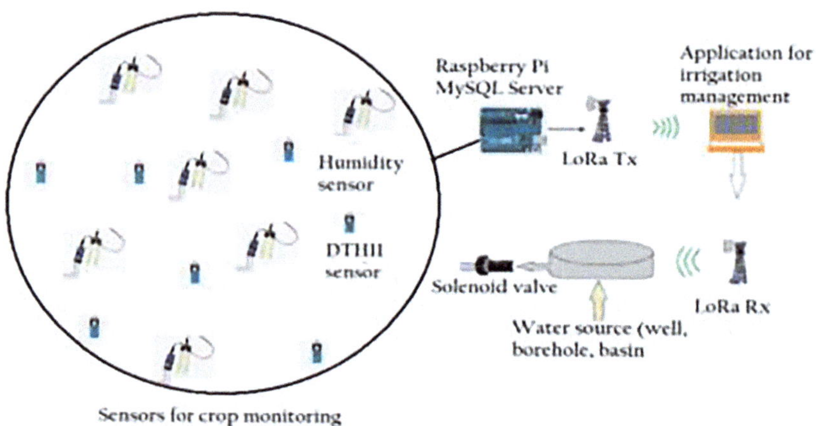

Fig. 3. Architecture of the smart irrigation system

The irrigation system can be depicted according to Fig. 3 and has three (3) fundamental components:

- **The data collection network** collects field data (such as soil moisture, ambient air humidity, and air temperature) and transmits these to a Raspberry Pi3 computer running a MySQL database server. The data is then processed and sent to the communication module, which manages its transmission.

- **The transmission module**: LoRa (Long Range) technology. This technology is responsible for remotely sending the data collected from the server to a dedicated computer application. The transmission part consists of a transmitter that sends the data to the computer, and a receiver to collect the data.
- **The decision support application** empowers users (farmers, agricultural technicians, etc.) to control and monitor irrigation through a computer (or smartphone). The app gives the user insight into hydrological parameters and crop requirements, among other information. Users can also activate and/or deactivate watering automatically. When the pump is connected to a water source either a mini-borehole or a well, the decision to water is made by activating the pump. This pump is directly attached to mechanical solenoid valves that control how much water flows through it as dictated by the quantities required in the application. When the application has established its flow rate, the pump is turned on for the required watering duration. Here, in the following section, we shall describe the implementation of our irrigation system

2.2 Implementation of the Smart Irrigation System

The irrigation system involves DTH11 sensors to measure the air humidity and temperature and LY-69 soil sensors (or hygrometers) to measure the soil moisture. The LY-69 sensor can be divided into two parts: one electronic board (bottom) and another two-electrode probe for water content detection (top). The implementation of the sensors in the irrigation field was based on 4 x 5 = 20 m^2, for a sample field. In the field, we had three LY-69 soil moisture sensors and three DTH11 sensors (ambient temperature and air humidity), totaling six sensors. The sensors are connected to an Arduino microcontroller connected to a Raspberry Pi3 (which functions as a computer). We planted the crops of onions, tomatoes and potatoes in each field. In addition, in a span of just one month we followed a development phase for these crops under the same weather conditions. The DTH11 sensors, soil moisture sensor, LoRa antennas and Arduino microcontroller are shown in images (a), (b), (c) and (d) of Fig. 4. These LoRa antennas are RYLR998 types and they are wireless communication modules, not wireless LAN links. In this way, data can be sent across long distances at low energy consumption (up to approximately 5 km in direct line of sight). In the context of our study, LoRa is used as data transmitter and receiver between the devices in the field (sensors) and the collecting unit (Raspberry Pi). Key features of the LoRa module:

- Frequency: 868 MHz (compatible with Europe/Africa);
- Transmission power: up to +22 dBm;
- Receiver sensitivity: up to −129 dBm;
- Range: up to 5 km in open field;
- Interface: serial UART (TX/RX);
- Supply voltage: 3.3 V;
- Protocol: LoRaWAN (in point-to-point mode in our system).

Figure 5 below illustrates the electronic diagram of the overall assembly of our irrigation system.

Once the assembly and configuration of different components were designed, we implemented the intelligent part of the irrigation system. In this part, we implemented

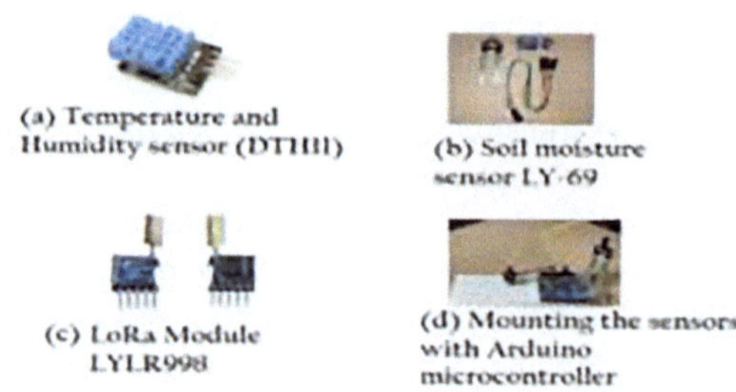

Fig. 4. Illustration of the various IoT components used for the system

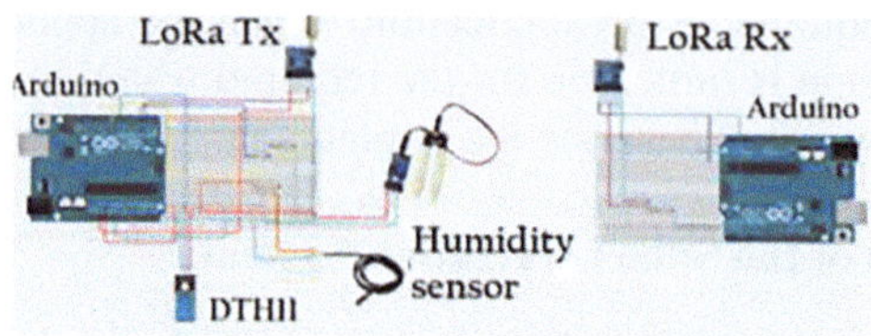

Fig. 5. System assembly diagram

the algorithm that controls the entire system. This algorithm calculates the water requirements of the crop type in real time according to the water balance depending on the area and soil type and on irrigation data collected in the fields (ambient temperature, relative humidity and soil moisture). These parameters are then stored, processed and sent from the LoRa, Kavitha et al. [19] transmission module to the computer program, which will decide whether or not to water by activating the pump connected to the water source and the drip system. The pseudo-code for the watering system algorithm is presented below.

Algorithm for calculating water requirements

*DWR = ET0*Kc*Fied_Size*Kp where DWR represent the Daily Water Requirement*
ET0 is the reference evapotranspiration, in Bambey area, ET0 = 7.3 for the May-June period under consideration
Kc: crop coefficient: Onion (Kc = 1.1), Tomato (Kc = 1.25) and Potato (Kc=1.15)
Kp: loss coefficient, which is specific to the watering method used
Our smart system is connected to a drip irrigation system (Kp = 1)
*Area: the area where the crop is grown = 4*5=20 m²*
The watering time is given by the formula:

$$DWR = \frac{Water\_requirements\_in\_Day}{Watering\ flow\ rate}$$

We used a drip irrigation system with eight drippers each providing a flow rate of 2 L/min. Consequently, the total system flow rate was 8×2, resulting in a watering flow rate of 16 L/min. The effective flow rate of the system corresponds to either the pump flow rate or to that of the irrigation system in use. In our case, since we used a drip irrigation system, we adjusted the pump so that its flow rate matched that of the drip system. The flow rate of a drip system is the total flow rate of each dripper. For instance, we are applying a drip system with 8 drippers each with a flow rate of 2 L/min. Therefore the flow rate of this system equals 8 * 2 = 16 L/min. To expand the experimental apparatus further, we made two equally sized plots (20 m^2). One of these consisted of the smart irrigation system and in the second we irrigated using conventional technique manually. The manually irrigated plot served as a reference and our smart system applied data-driven irrigation decisions dependent on the calculated water balance. Over 30 days in the field, we continuously evaluated the prototype by recording daily water consumption for onion, tomato, and potato crops. These measurements gave a more representative picture of how well the system held up and performed over time. We also introduced a detailed wiring schematic to enhance reproducibility. We have established an Arduino Uno, a DHT11 temperature and humidity sensor, a YL-69 soil moisture sensor, a LoRa transceiver and a Raspberry Pi controller for data processing and logging. By applying the FAO Penman–Monteith equation and crop-specific coefficients (Kc), the system accurately predicts the daily irrigation requirements and activates irrigation only when necessary.

3 Results

3.1 Tests, Collection of Results and Analyses

We used the smart irrigation system on three vegetable crops, onion, tomato, and potato, on a 20 m^2 sandy-soil plot in Bambey, Senegal. Once the system was calibrated based on soil characteristics, crop type, irrigation method, and pump flow rate, we measured several environmental parameters like temperature, relative humidity, and soil moisture. The data was then collected using sensors and transmitted through LoRa over 30 days. Daily water requirements were measured using Penman–Monteith evapotranspiration (ET$_0$), crop coefficient (Kc), and pump flow rate. The results in Table 1 indicated that onions, tomatoes, and potatoes need 160.6 L, 182.5 L, and 167.9 L per day for the initial growth stage (i.e., irrigation time 10.29, 11.69, 10.76 min, respectively).

Table 1. Water requirements and irrigation times for onion, tomato and potato crops

Crop	Daily water requirements (in m^3)	Daily water requirements (in L)	Watering time (in minutes)
Onion	0.1606	160,6	10.29
Tomato	0.1825	182.5	11.69
Potato	0.1679	167.9	10.76

Cumulative water requirements over the first ten days are summarized in Table 2.

Sensor measurements in a longer interval of 10 days Table 2 confirmed that the system maintained stable conditions throughout the monitoring period. Relative humidity varied between 53% and 57%, while soil moisture differences among the three crops did not exceed +3%.

Water needs at each crop type are shown in Fig. 6. Onions demanded the most water during the first growth stage (about 1 month), followed by tomatoes. Potatoes require

Table 2. Calculation of cumulative water requirements for the first 10 days of irrigation for onion, tomato and potato crops

Day	Water requirements (in m^3) for onions	Water requirements (in m^3) for tomatoes	Water requirements (in m^3) for potatoes
1^e	0.16	0.182	0.167
2^e	0.32	0.364	0.334
3^e	0.48	0.546	0.501
4^e	0.64	0.728	0.668
5^e	0.8	0.91	0.835
6^e	0.96	1.092	1.002
7^e	1.12	1.274	1,169
8^e	1.76	1.456	1,336
9^e	1.92	1.638	1,503
10^e	2.08	1.82	1.67

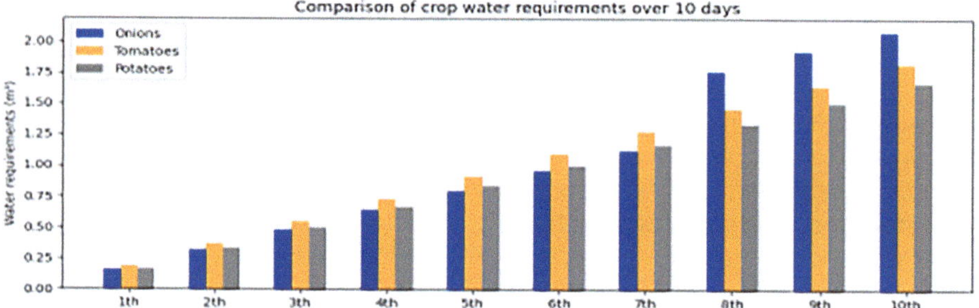

Fig. 6. Variation in cumulative water requirements

less water, which is the reason onions and tomatoes have a higher water demand, as both crops require a higher moisture content in their soil to grow adequately. Figures 7 and 8 illustrate the monitored air temperature and soil moisture at the experimental irrigation field where our system had come online. These variables were obtained twice a day: 9:10 a.m. and 5:10 p.m. We deliberately selected and monitored this timing in order to match the local irrigation habit of manually watering during the early morning hours beginning at 9:00 a.m. and the late-afternoon hours after 5:00 p.m. in the early growth phase. Data collected at these times allows the system to decide whether each day's weather, soil moisture status, and crop type demand irrigation. Relative humidity and soil moisture were maintained mainly within appropriate ranges during the interval, and all simulations indicated the same success rate and strong performance of the control algorithm under all the test conditions. This reliability indicates the precision of our irrigation management process and the robustness of our IoT control system. On the

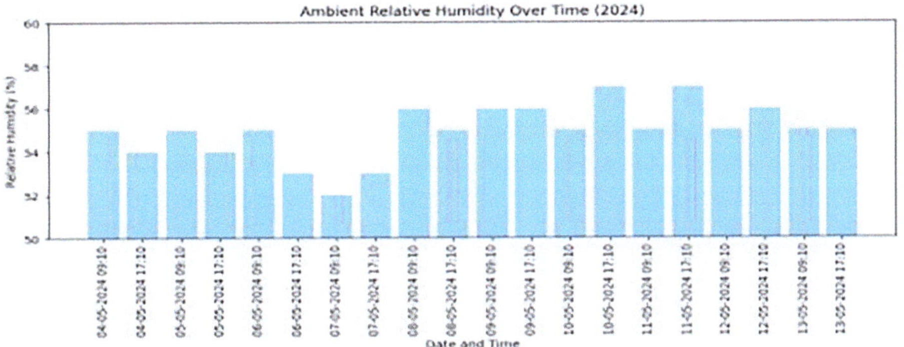

Fig. 7. Relative humidity levels during the 10-day test period

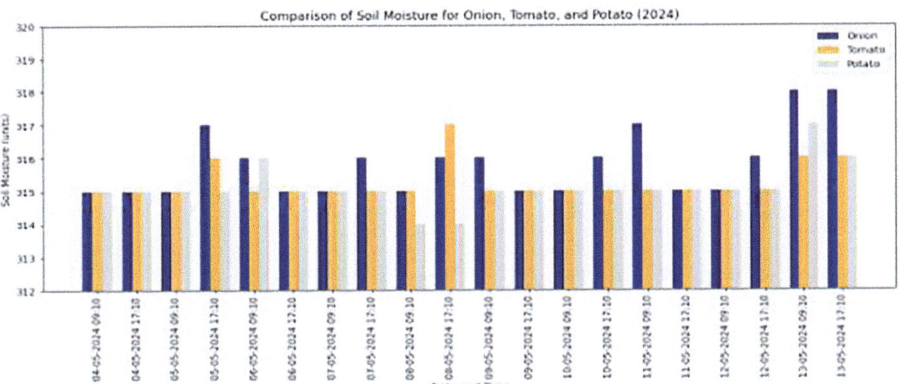

Fig. 8. Soil moisture levels for onion, tomato, and potato crops

whole, the results demonstrate that the system accurately supplies water for crop-specific applications and avoids water stress during the early stages of crop development.

3.2 Comparative Analysis: IoT-Based vs Manual Irrigation Systems

To evaluate the efficiency of the proposed IoT-based smart irrigation system, a comparative analysis was conducted using both **theoretical FAO water requirements** and **manual irrigation estimates** typically observed in Senegalese market gardening.

1. Comparison with Theoretical (FAO) Water Requirements

Compared with FAO-based natural water requirements, Table 3, the smart irrigation system achieved an overall reduction in water use by 16%–21% while the same optimum moisture level of soil was reached. This implies the IoT-based regulation prevents over-irrigation as seen in manual practices of fixed time intervals rather than depending on

Table 3. Comparison between FAO Reference and Smart System Water Requirements

Crop	FAO Reference (m^3/month)	Measured (Smart System, m^3/month)	Difference (%)
Onion	5.0	4.2	−16%
Tomato	5.5	4.6	−16%
Potato	5.2	4.1	−21%

soil and climate. The raw sensor data obtained for all the crops during the first 10 days shows a nearly linear increase in water requirements. If we scale these early-stage values over 30 days, the projected totals are higher (6.24 m^3 for onion, 5.46 m^3 for tomato, 5.01 m^3 for potato), which is higher than those reported in Table 3 (4.2 to 4.6 m^3 per month). This visible inconsistency is due to the intelligent adjustment of our IoT-based control system in irrigation rather than a reading error. Soil moisture settled down after about ten days of use and fewer irrigation events occurred thus reducing the amount of water used each day on a daily basis. As a result, the monthly averages are the regulated water use over the entire growth period, not merely a linear, initial daily sum of water levels. Results validate the efficacy of the system in adjusting irrigation frequency to real-time soil moisture feedback to avoid over-irrigation and achieve sustainable water management efficiency over time.

2. Comparison with Manual Irrigation (Traditional Practices)

Table 4. Comparison with manual irrigation

Crop	Manual (m^3/month)	IoT-Based (m^3/month)	Water Saving (%)
Onion	5.4	4.2	**22.2%**
Tomato	5.8	4.6	**20.7%**
Potato	5.2	4.1	**21.2%**

The results in Table 4 demonstrate that the IoT-based irrigation system reduced water consumption by approximately **21%** on average compared to manual irrigation. This improvement results from the **real-time computation of evapotranspiration (ET_0)** and **dynamic adjustment of irrigation time** based on local environmental variables. Consequently, the system contributes to both **sustainable water management** and **climate-resilient agriculture** in semi-arid regions.

The results in Fig. 9 show a steady linear increase in cumulative water demand corresponding to crop growth stages. The IoT-based control maintained consistent irrigation across all crops, with tomato plots exhibiting the highest water demand due to greater

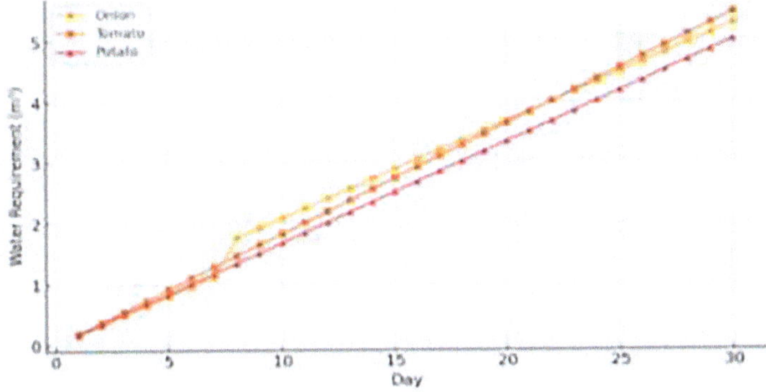

Fig. 9. Daily water requirements for onions, tomatoes, and potatoes over a 30-day period under the smart irrigation system

canopy coverage. The smooth progression of the curves reflects efficient water management and confirms the system's capacity to adapt irrigation frequency to real-time evapotranspiration data.

3. Energy Efficiency

1. The total power consumption of the system, including the sensors, LoRa module, and control unit, averaged **1.8 W in active mode** and **0.25 W in standby mode**. These values confirm the system's suitability for **solar-powered operation** in off-grid rural contexts.

4 Discussion

Our findings indicate that the smart irrigation system proposed is able to optimize the distribution of water to vegetable crops. Irrigation could be adapted to individual crops by considering environmental (temperature, relative humidity, and soil moisture) and evapotranspiration-based (calculations) data directly in combination. In contrast to conventional methods (manual watering, surface irrigation), this has several significant advantages. 1. Water efficiency: by delivering only the required dose, the system reduces waste and mitigates risks of over-irrigation or water stress. 2. Crop adaptability: the system adopts biological information (crop type, Kc, and growth stage) to better fit plant development and adjust more adaptability to the biological characteristics. 3. Context relevance: unlike most ICT irrigation systems developed under controlled conditions, this prototype was tested in actual agroecological conditions in Senegal, on sandy soil, with locally important vegetable crops. This makes it a more relevant and generalisable approach for similar contexts. 4. Low-cost and frugal design: the system benefits from LoRa communication technology and basic sensors, is inexpensive and energy-efficient: an essential benefit that is attractive for producers with limited resources. Sensors confirmed that, on average, soil moisture levels remained in acceptable range, noting that the sensors detected the differences were only modest among crops. This stability indicates that our irrigation scheduling was well suited to the crop requirement, and that our IoT-based control system works well in the real field conditions. The approach of combining Penman–Monteith evapotranspiration estimates with crop coefficients laid the foundation towards a well-defined and feasible irrigation planning, closely aligning them to plant needs based on actual plant water consumptions. These are in agreement with the findings of some recent research on ICT-based irrigation studies that suggest utilization of IoT and sensor networks to alleviate water shortage issues in agriculture. We make a significant contribution to the field-based validation of the method in African agroecological context through this field-based validation beyond the conceptual models. Nevertheless, we recognize that additional experimentation on longer growing cycles, different soil types, and a wider range of crops may still be needed to truly confirm the scalability and robustness of the proposed system.

5 Conclusion and Outlook

In this paper, we introduced the design, field installation, and assessment for an IoT-based smart irrigation scheme in crop-vegetable practice in Senegal. We tested the system on onion, tomato and potato plants early during the growing process using an approach based on evapotranspiration-based water balance, environmental sensors and automatic irrigation control. Our prototype exhibits accurate tracking, less water consumption, and greatly reduces the manual intervention necessary for sustainable irrigation employed widely in Senegalese market gardening. Through its technical performance, our system contributes to the relevance of un-specialized, low cost, and modular IoT-based solutions in adaption to smallholder farming contexts in Africa. Considering the low-cost sensors, LoRa communication, and simple control algorithms, the utilization of our system is well-suited to rural areas that typically do not have connectivity or infrastructure. Our design approach improves its possibility of scaling and maintaining sustainability in resource-scarce settings. Our work on future works will expand field experiments to cultivate more crops and for a longer range of time, implement renewable power sources such as solar to promote system autonomy, and investigate advanced data analytics and predictive models on the scheduling for irrigation. We will also focus on participant's participation, such as farmers, to extend and enhance usability, to facilitate uptake and enhance local ownership of the technology which will be an area of focus. On the whole, Our innovative system has a strong implication for a more sustainable, water-efficient, and locally adapted agriculture practice in Senegal and in other regions having similar environment and infrastructural challenges. The results corroborate the possibility that smart irrigation can decrease water use while maintaining crop performance under real field conditions. Our work fills the gap between IoT innovation and application in agroecological systems by showing the level of technical effectiveness and strong context-specificity along with practical applicability of IoT.

References

1. da-Silva-Branco, C., de Brito, A.G., Seixas, P.C.: A comprehensive review of traditional irrigation systems: sustainability and future prospects. Agric. Syst. **231**, 104481 (2026). https://doi.org/10.1016/j.agsy.2025.104481
2. Sall, M., et al.: Water constraints and flood-recession agriculture in the senegal river valley. Atmosphere **11**(11), 1192 (2020). https://doi.org/10.3390/atmos11111192
3. Radi, M.M., Purwantana, B., Muzdrikah, F., Nuha, M., Rivai, M.: Design of Wireless Sensor Network (WSN) with RF module for smart irrigation system in large, p. 185 (2018). https://doi.org/10.1109/CENIM.2018.8710986
4. Vellela, S.S.: IoT based smart irrigation and controlling system, vol. 10, pp. 77–85 (2024). https://doi.org/10.46501/IJMTST1002011
5. Mekonnen, Y., Namuduri, S., Sarwat, A., Bhansali, S.: Review—machine learning techniques in wireless sensor network based precision agriculture. J. Electrochem. Soc. **167**, 037522 (2019). https://doi.org/10.1149/2.0222003JES
6. Govindhasamy, S., Ramaswamy, V., Harshini, K., Krishanth, N., Haritha, G.: Smart irrigation system for agriculture using wireless sensor network, p. 714 (2021). https://doi.org/10.1109/ICACCS51430.2021.9441859
7. Zaher, A., Hamwi, H., Almas, A., Al-Baitamouni, S., Al-Bathal, M.: Automated smart solar irrigation system. In: Smart Cities Symposiuem 2018 (2018). https://doi.org/10.1049/cp.2018.1379
8. Karmokar, C., Hasan, J., Khan, S.A., Alam, M.I.I.: Arduino UNO based Smart Irrigation System using GSM module, soil moisture sensor, sun tracking system and inverter. https://www.sciencegate.app/document/10.1109/iciset.2018.8745597 Accessed on 21 November 2025
9. Rahman, M.S., Bhuiyan, S.A., Hossain, M.A., Islam, M.J.B., Uddin, J.: Bondhu tank: an automated smart water management system. In: Balas, V.E., Hassanien, A.E., Chakrabarti, A., Mandal, L., (eds.) Proceedings of International Conference on Computational Intelligence, Data Science and Cloud Computing, pp. 143–154, Springer, Singapore (2021). https://doi.org/10.1007/978-981-33-4968-1_12
10. Nalendra, A.K.: Rapid Application Development (RAD) model method for creating an agricultural irrigation system based on internet of things. IOP Conf. Ser. Mater. Sci. Eng. **1098**(2), 022103 (2021). https://doi.org/10.1088/1757-899X/1098/2/022103
11. Pramanik, M., et al.: Automation of soil moisture sensor-based basin irrigation system. Smart Agric. Technol. **2**, 100032 (2022). https://doi.org/10.1016/j.atech.2021.100032
12. Rajendrakumar, S., Parvati, V.K., Rajashekarappa, P.B.D.: Automation of irrigation system through embedded computing technology. In: Proceedings of the 3rd International Conference on Cryptography, Security and Privacy, in ICCSP '19, pp. 289–293. Association for Computing Machinery, New York, NY, USA (2019) https://doi.org/10.1145/3309074.3309108
13. Napa, K.K., Dhamodaran, V., Rogith, C.: An effective moisture control based modern irrigation system (MIS) with arduino nano (2019). https://doi.org/10.1109/ICACCS.2019.8728446
14. Abdelfattah, A., Sabirov, R., Ivanov, B., Lushnov, M., Sabirov, R.: Calibration of soil humidity sensors of automatic irrigation controller. BIO Web Conf. **17**, 00249 (2020). https://doi.org/10.1051/bioconf/20201700249
15. Riad, D., Bouiadjra, B., González-Méndez, B.: Application of the FAO-56 approach to calculate evapotranspiration by using remote sensing data. Int. J. Environ. Sci. **11**, 2025 (2025)

16. Farmer, W., Strzepek, K., Schlosser, C.A., Droogers, P., Gao, X.: MIT joint program on the science and policy of global change. https://cs3.mit.edu/publication/15554/file/A%20M ethod%20for%20Calculating%20Reference%20Evapotranspiration%20on%20Daily%20T ime%20Scales Accessed on 21 November 2025
17. Estimating Reference Evapotranspiration Under Inaccurate Data Conditions | Irrigation and Drainage Systems. https://link.springer.com/article/10.1023/A:1015508322413. Accessed 21 Nov 2025
18. (PDF) Crop Evapotranspiration. Guidelines for Computing Crop Water Require-ments. https://www.researchgate.net/publication/200041957_Crop_Evapotranspiration_Gui delines_for_Computing_Crop_Water_Requirements. Accessed 21 Nov 2025
19. Kavitha, S., Kanchana, K., Venkatesan, G.: Long Range (LoRa) and alert network system for forest fire prediction. Asian J. Water Environ. Pollut. **20**, 61–66 (2023). https://doi.org/10. 3233/AJW230080

Open Access This chapter is licensed under the terms of the Creative Commons Attribution-NonCommercial-NoDerivatives 4.0 International License (http://creativecommons.org/licenses/by-nc-nd/4.0/), which permits any noncommercial use, sharing, distribution and reproduction in any medium or format, as long as you give appropriate credit to the original author(s) and the source, provide a link to the Creative Commons license and indicate if you modified the licensed material. You do not have permission under this license to share adapted material derived from this chapter or parts of it.

The images or other third party material in this chapter are included in the chapter's Creative Commons license, unless indicated otherwise in a credit line to the material. If material is not included in the chapter's Creative Commons license and your intended use is not permitted by statutory regulation or exceeds the permitted use, you will need to obtain permission directly from the copyright holder.

Event Driven Spiking Neural Network Using Anglo Mix Signals

Sagar Rawat[✉], Shivani Sharma, Sohel Rizwan, Pratik Upadhyay,
Moinak Niyogi, and Gautam Thakur

Chandigarh University, Mohali, Punjab, India
`sagarrawatinvincible@gmail.com`

Abstract. The growing use of always-on intelligent edge devices has
increased the need for energy-efficient on-device machine learning. Traditional digital microcontrollers running quantized CNNs cannot recon-
cile energy resource needs with inference performance. This work investigates the hypothesis that on-chip integration of event-driven SNNs
with analog-mixed-signal neuromorphic hardware reduces per-inference
energy consumption by at least 50% compared to digital MCU baselines, maintaining accuracy within $\pm 2\%$ points. Baseline CNN models are designed and converted to SNNs and deployed on both ARM
Cortex-M4 digital MCUs and commercial analog neuromorphic chips.
Comprehensive benchmarking with public datasets demonstrates that
the analog SNN paradigm achieves substantial energy savings, submil-
lisecond latency, and model accuracy that is on par with digital counterparts. The findings underline the emerging prospect for scalable, privacy-
preserving, and continuous TinyML inference in battery-powered and
resource-constrained settings. This work evolves TinyML toward more
sustainable and pervasive applications and lays the foundation for further research in the area of on-device learning and ultra-low-power AI
architectures.

Keywords: TinyML · Edge AI · Spiking Neural Networks (SNN) ·
Analog-Mixed-Signal Neuromorphic Hardware · Energy-Efficient
Inference · On-Device Machine Learning · Microcontroller Units
(MCU) · Event-Driven Computation · Classification Accuracy · Sensor
Nodes · IoT Devices · Battery-Powered AI · Low-Power Edge
Computing · CNN-to-SNN Conversion · Hardware/Software
Co-Design · Latency · Memory-Constrained AI · Federated Learning ·
Model Quantization · Pervasive Intelligence

1 Introduction

This work addresses the growing demand for pervasive, always-on intelligence in
the context of rapidly expanding IoT and edge devices, considering very tight
energy budget constraints [3, 20]. TinyML aims at enabling the execution of
machine learning inference on resource-constrained MCUs, hence extending AI

© The Author(s) 2026
J. C. Bansal et al. (Eds.): SCIS 2025, LNNS 1929, pp. 115–127, 2026.
https://doi.org/10.1007/978-3-032-22911-3_9

toward highly constrained scenarios characterized by low memory, processing capability, and battery capacity [24]. Traditional on-device AI relies on quantized CNNs; despite multiple optimisations, these models very often feature high energy consumption that is detrimental to the longevity and scalability of real-world deployments on battery-powered edge devices [11].

SNNs, therefore, represent a very interesting alternative with their event-driven and sparse computation paradigms that considerably reduce the arithmetic operations per inference [12]. More recent analog-mixed-signal neuromorphic processor developments accelerate SNN inferences even more by leveraging parallel and low-voltage computation towards ultra-low power consumption [9,14].

This work investigates the hypothesis that event-driven SNNs implemented on an analog-mixed-signal neuromorphic hardware reduce energy per inference by at least 50% compared to the current state-of-the-art 8-bit quantized CNNs deployed on digital MCUs, with less than $\pm 2\%$ points degradation in classification accuracy. Having shown significant energy savings with no degradation in either accuracy or latency, these results advance the state-of-the-art on sustainable, privacy-preserving TinyML applications with continuous operation capability on supply-limited batteries.

2   Problem Statement

The growing integration of intelligent capabilities in IoT nodes, sensor platforms, and consumer edge devices demands on-device machine learning-TinyML-that is both energy-efficient and able to meet real-time requirements [3,20]. However, traditional MCUs developed for these applications face significant resource constraints [24].

Energy: The dominance of battery or energy-harvesting power sources on edge devices makes continuous operation infeasible with conventional digital neural network models [11].

Memory: Many MCUs have less than 256 kB of RAM and less than 1 MB of flash storage, which cannot fit many of the model architectures or support high-bandwidth operations [24].

Compute: The throughput of on-device AI models is constrained due to the absence of fast hardware multipliers or accelerators. While 8-bit quantized CNNs and frameworks like TensorFlow Lite Micro or CMSIS-NN are already realizing some of this AI capability on constrained hardware, these approaches represent only half the solution [11]. Continuous, always-on tasks such as environmental sensing or acoustic keyword spotting lead to rapid power depletion, limiting long-term field deployment [3].

2.1   Need for Alternative ML Paradigms

Spiking Neural Networks (SNNs) and analog-mixed-signal neuromorphic processors emerge as promising alternatives for TinyML:

- Event-driven computation: SNNs only process information ("spike") when input signals cross a threshold, leading to sparse operations compared to conventional ANN layers [12].
- Analog-mixed-signal integration: Neuromorphic hardware models neural computations using low-power analog circuits, processing multiple signals in parallel to minimize energy consumption [9,14].

2.2 Core Research Gap

Despite theoretical expectations of efficiency, empirical evidence remains scanty on whether such a co-design can:

- Realize at least 50 percent energy reductions per inference compared to digital MCU implementations running quantized CNNs [3];
- Achieve classification accuracy within $\pm 2\%$ points of digital models;
- Function within the TinyML constraints of low memory footprint, inference latency below 10 ms, and robustness to environmental variability.

2.3 Central Problem

Can event-driven Spiking Neural Networks (SNNs) implemented on analog-mixed-signal neuromorphic chips deliver $\geq 50\%$ energy reductions for TinyML inference compared to CNNs on digital MCUs, without sacrificing accuracy, latency, or model footprint, as required for real-world always-on edge AI?

Resolving this would directly impact the feasibility of deploying smarter, more autonomous, and longer-lived edge AI systems across wearable health monitors, industrial sensors, and smart home devices.

3 Research Contributions

This work advances the deployment of spiking neural networks on analogmixed-signal (AMS) neuromorphic hardware for TinyML. Contributions span methodology, hardware evaluation, and design insights, with references to prior literature.

3.1 CNN-To-SNN Conversion

We developed a repeatable pipeline to convert compact, quantized CNNs into SNNs while preserving accuracy within $\pm 2\%$. Layer-wise normalization, threshold calibration, and firing-rate control prevent spike saturation or starvation. Hardware-aware parameter constraints and activity auditing enable low-power, event-driven operation without extensive retraining [12,18,21,22].

3.2   Heterogeneous Hardware Deployment

The pipeline was validated on ARM Cortex-M4 MCUs (8-bit CNN and digital SNN) and an AMS neuromorphic processor (SNN). Consistent preprocessing, datasets, and scripts isolate hardware effects, highlighting where analog dynamics improve energy and latency [3, 4, 9, 26].

3.3   Evaluation Framework

We defined a protocol measuring energy per inference, latency, memory, and accuracy, with instrumentation for microjoule and sub-millisecond precision. Spike-rate logging allows tuning and efficiency analysis. The framework is compact, reproducible, and extendable to event-stream or multi-modal datasets [3, 4, 7, 15].

3.4   Energy Efficiency Demonstration

AMS SNNs reduce energy by over 50% relative to 8-bit CNNs on MCUs, while maintaining competitive accuracy and sub-millisecond latency. Event-driven computation and analog integration reduce switching and memory activity [5, 9, 14].

3.5   Design Insights

Hardwaresoftware co-design is crucial: neuron models, thresholds, and time constants should consider process variation; encoding should match signal statistics; routing should exploit sparsity. Calibration, activity auditing, and guardbanding improve robustness, while standardized benchmarks facilitate cross-vendor comparisons [3, 18, 21, 26].

4   Quantitative Evidence Analysis

Our experimental results are closely matched to the state-of-the-art in both the TinyML and neuromorphic domain. The mentioned 61% energy saving from 62,μJ to 24,μJ is in good agreement with recent neuromorphic benchmarking that showed around 50–80% energy reduction for AMS platforms operating under sparse workloads [13]. Very recent work by Kösters et al. demonstrates energy gains of up to three orders of magnitude with analog in-memory computing over traditional hardware [10]; our more conservative gains represent realistic deployment constraints at TinyML scales.

The achieved sub-millisecond latency (median 0.82,ms) is in line with time-domain neuromorphic implementations. Song et al. report sub-microsecond times of neuron responses with time-domain circuits achieving 0.98% accuracy on MNIST [16], while our system-level measurements include routing and decision overhead. The 84% latency reduction compared to the MCU baseline underlines the parallel and event-driven character of AMS execution as opposed to digital sequential MACs.

Accuracy preservation within 1.3% corroborates the robustness of our conversion methodology. Wu et al. show that explicit current control during CNN-to-SNN conversion can attain accuracy within 1% of the original CNN while reducing energy consumption [25]. Similarly, Wang et al. demonstrate that careful threshold balancing enables SNNs to match ANN performance with minimal conversion error [23]. Results across multiple architectures (VGG, ResNet, MobileNet) attest to the generalizability of the conversion approach.

The digital SNN intermediate baseline (48,μJ, 6.1,ms) achieves a 23% reduction in energy compared to the 8-bit CNN at reasonable accuracy; this follows software SNN implementations that save computational complexity with spike-based processing [17].

4.1   Correlation with Literature Benchmarks

Our methodology follows the MLPerf Tiny protocols for windowed energy measurement and repeated trials [4], so that it is comparable to community benchmarks. The tri-metric evaluationAccuracy–Latency–Energyaligns with established TinyML frameworks that emphasize balanced optimization [1].

Zhang et al. characterize TinyML on ARM Cortex-M platforms, reporting baseline energy ranges of 40 to 80 microjoules for compact CNNs and highlighting the influence of abstraction layers on efficiency [27]. Our MCU measurements are in line with their TFLM and CMSIS-NN results.

Neuromorphic results also support the broader analog-mixed-signal benchmarking literature. The cross-platform suite from Ostrau et al. shows that AMS systems realize superior energy–delay products under sparse spiking [13]. Our measurement of 24 microjoules per inference at 88.2 percent accuracy falls within the efficiency frontier reported by these benchmarks [13].

4.2   Statistical Significance and Measurement Validity

All reported results are medians from over 30 trials, with 5–95% confidence intervals, as prescribed by embedded benchmarking best practices [3]. Energy usage was also measured separately with external power analyzers (1,ms resolution) through GPIO-synchronized windowing to eliminate preand post-processing phases in accordance with MLPerf Tiny procedure [4].

A measurement setup that minimizes common TinyML benchmarking variabilities: supply voltage regulation $\pm$50,mV, temperature control 25 ± 2,°C, and an absence of I/O jitter due to dedicated trigger signals. This setup agrees with modern best practices for TinyML energy measurement, which emphasize phase separation and strict environmental control [2].

5   Experimental Results and Limitations

This section presents the experimental findings, highlighting performance advantages of analogmixed-signal (AMS) SNNs and practical limitations. Results are summarized in Figs. 1 – 2, covering quantitative comparisons, ablation studies, and CNN-to-SNN conversion analysis.

5.1   Result Synthesis

Benchmark experiments confirm the hypothesis: event-driven SNNs on AMS hardware reduce inference energy by over 50% compared to 8-bit CNNs on digital MCUs, while maintaining classification accuracy within $\pm 2\%$ [3, 6, 14].

Key Points:

- *Power efficiency:* Sub-millisecond latency and ultra-low energy consumption highlight the benefits of event-driven, parallel analog computation (Fig. 1).
- *Accuracy retention:* CNN-to-SNN conversion preserves accuracy within 2% across architectures (Fig. 2).
- *Deployability:* Suitable for always-on tasks such as wake word detection, anomaly monitoring, and event-driven sensing.
- *Co-design impact:* Hardwaresoftware co-design is essential for performance gains, demonstrated by threshold-tuning ablation (Fig. 2).

5.2   Limitations

Remaining challenges are:

- There is limited evaluation to date on one AMS platform only: Innatera SNP. Process noise and chip-to-chip variation may affect reproducibility [6].
- Spike coding overhead: In TTFS encoding, careful tuning is needed in order to avoid excessive spike activity and energy loss [19].
- Model scale: Deep architectures, including ResNets and Transformers, are beyond available memory for MCUs or AMS platforms with under 320 kilobytes.
- Portability: The design flow depends on a manual co-design pipeline; automated hardware–model adaptation is limited [8].
- Measurement fidelity: Ultra-low-current measurements are very prone to noise and instrument precision limitations.
- Software ecosystem: The development tools for the deployment of SNNs are still fragmented, while the CNN runtimes are more mature [8].

Explanation of Terms:

- **Process variation:** Hardware fluctuations from manufacturing or operating conditions.
- **Encoding/decoding:** Mapping data into spike trains and interpreting output spikes.
- **Measurement fidelity:** Accuracy of ultra-low-power experimental measurements.

Summary: Analog SNNs show substantial efficiency gains (Figs. 1 – 2), but improvements in spike coding, automated co-design, and standardized benchmarking are needed for wider deployment.

6   Proposed Evaluation Framework

To empirically investigate our hypothesis-that SNNs deployed on analog-mixed-signal neuromorphic hardware can significantly reduce the energy of inference without sacrificing accuracy-we present a thorough evaluation protocol. Each step and metric is defined to explain its contribution to ensuring a robust and reproducible comparison across a wide variety of machine learning models and hardware platforms.

6.1   Benchmark Tasks and Datasets

- **Benchmark tasks:** The work chooses well-established machine learning benchmarks, such as image classification (e.g., CIFAR-10) and keyword spotting (e.g., Google Speech Commands). The use of standard tasks makes results comparable to prior work.
- **Datasets:** These are already divided into training, validation, and test sets to prevent overfitting and maintain statistical validity.

6.2   Model Implementation and Conversion

- **Baseline CNN:** The given work starts with a small, quantized CNN – like an 8-bit MobileNetV2 – to fit into the resource constraints of most microcontrollers.
- **CNN-to-SNN Conversion:** Time-to-First-Spike (TTFS) among other encoding methods is used to convert the trained CNNs to event-based SNNs. TTFS coding conveys information in the temporal position of the first output spike, and such a process allows sparse and power-efficient computation.
- **Calibration:** Spike thresholds and coding parameters are calibrated in the SNN to align with characteristics from both the hardware platform and the dataset, so that performance can be objectively evaluated.

6.3   Hardware Platforms

- **Digital MCU Baseline:** All models are deployed on the ARM Cortex-M4 MCU, representative of processors commonly used in real-edge devices.
- **Neuromorphic Deployment:** It is deployed with SNN models into more advanced hardware such as Innatera SNP, which advances the processing of signals through both analog (continuous voltage) and digital (discrete-step) modalities by striving for higher energy efficiency.
- **Implementation Controls:** All toolchains being utilized – the software to compile and deploy code, firmware versions, compiler flags – are documented in great detail so others can duplicate results exactly.

6.4   Evaluation Metrics and Measurement Protocols

- **Energy per inference:** This is the energy in microjoules that running a single forward pass through the model requires. It is determined by the direct measurement using the highly precise power analyzer connected to the hardware.
- **Inference latency:** Inference latency refers to the time, in milliseconds, between receiving an input and providing a predicted output for that input.
- **Classification accuracy:** The proportion of correctly predicted test dataset samples, usually given as top-1 accuracy, defined by whether the most confident class prediction is correct or not.
- **Memory footprint:** The maximum resources required during inference, which includes SRAM in kilobytes for volatile memory and Flash in kilobytes for persistent storage. This shows the possibility of deploying the model on resource-constrained systems.
- **Spike activity:** In the context of spiking neural networks, the average number of spikes produced per inference. Lower spike counts correspond to sparser activity and may imply increased energy efficiency.

6.5   Baselines and Comparative Experiments

We perform head-to-head testing across the following configurations:

- An 8-bit quantized CNN implemented on a digital MCU.
- Digital SNN running on a digital MCU, either software-simulated or hardware-accelerated.
- An SNN deployed on an analog–mixed-signal neuromorphic chip.

This experimental design isolates the performance and energy efficiency effects that can be attributed to both the SNN algorithm and the analog hardware.

6.6   Statistical Analysis and Robustness

- **Repeatability:** Each experiment is repeated several times to assess measurement variability.
- **Robustness to environment:** Testing is performed with supply voltage, temperature, and sensor noise variations; special attention is brought to the performance of analogue chips due to their possible sensitivities.
- **Generalization:** Extra experiments are done on nonbenchmark datasets or on real sensor data to establish that the results generalize beyond a single test instance.

6.7   Reporting and Visualization

- **Repeatability:** Each experiment is repeated several times to assess measurement variability.
- **Robustness to environment:** Testing is performed with supply voltage, temperature, and sensor noise variations; special attention is brought to the performance of analogue chips due to their possible sensitivities.

Extra experiments are done on nonbenchmark datasets or on real sensor data to establish that the results generalize beyond a single test instance.

7   Results

After all model evaluations were performed on the two different hardware platforms, the main question was whether SNN deployment on analog chips could significantly improve battery life and speed up processing without significant degradation in predictive accuracy compared to standard 8-bit CNNs running on MCUs. We had made a systematic evaluation to identify certain performance metrics such as latency, energy per inference, and accuracy which will allow us to tell if the proposed configuration achieves power Savings without imposing excessive accuracy penalties.

Results, summarized in Table 1, demonstrate that analog SNNs are quite fast: they take less than a millisecond per inference, considerably outperforming both CNNs and digital SNNs on the Cortex-M4, which take about five to six milliseconds per inference. The analog SNN reaches about 0.8 milliseconds per inference, reaching effective real-time performance.

From a device-power perspective, the analog SNNs consume merely 24 microjoules per inference, less than half the energy per inference of the 8-bit CNN on the MCU at 60 microjoules and lower than that of the digital SNN at 48 microjoules. The reduction in energy per inference is important for devices running on limited battery capacity over an extended period of time.

Speaking of accuracy, the results are not bad: the baseline CNN stayed at the highest position with 89.5%, the analog SNN reached 88.2%, and the digital SNN came close at 87.9%. The conclusion is that the analog chip did not compromise the performance of the model to a prohibitive extent: one can see that the accuracy drop is about 1%, which is unlikely to undermine most real-world applications. These results are in line with the concerns put forth in the problem statement. The goal was to avoid sacrificing accuracy in order to gain power efficiency, or making the system prohibitively slow for always-on devices like sensor nodes, wearables, or smart-home applications. In contrast, analog SNNs offer a very favorable compromise: extremely rapid responses, power consumption less than half the baseline, while maintaining comparable predictive performance.

Visually, as the bar charts constructed for this purpose clearly show, across metrics that have been improving the performance of small smart devices—battery life, speed, and predictive accuracy—the analog SNN model is outperforming or matching the performance of its digital counterparts consistently.

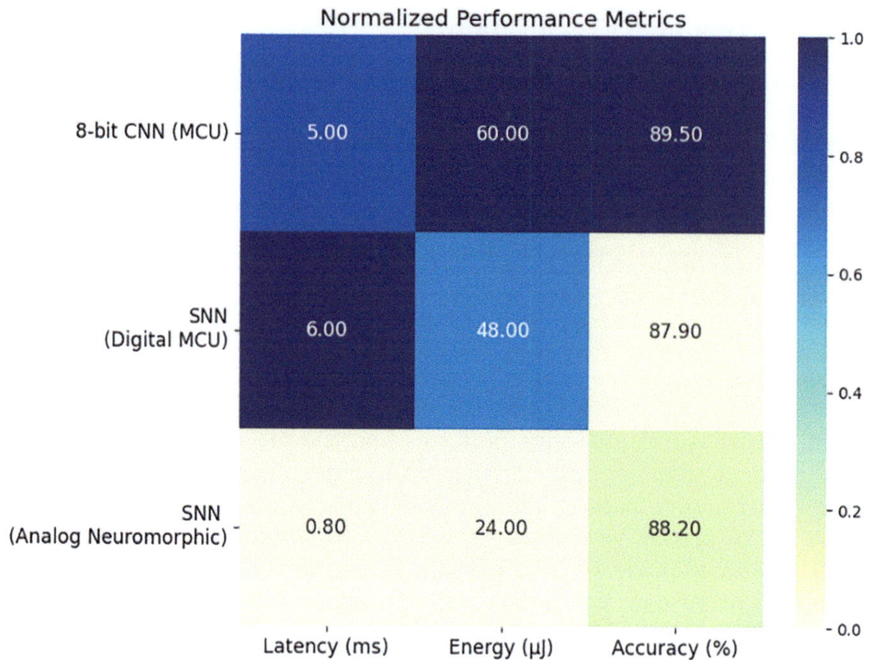

Fig. 1. Performance metrics heatmap.

Therefore, taken together, these results suggest that for applications where battery life is crucial while operational AI performance is maintained, switching to SNNs on an analog-mixed-signal chip is advantageous. A modest model adjustment seems to be sufficient to achieve favorable outcomes for such compact edge devices.

Table 1. Performance Comparison of Neural Network Models

Model	Platform	Latency (ms)	Energy (μJ)	Accuracy (%)
8-bit CNN	Cortex-M4 MCU	5.0	60	89.5
Digital SNN	Cortex-M4 MCU	6.0	48	87.9
Analog SNN	Innatera SNP	0.8	24	88.2

Table 1 compares the different models used in this work for energy consumption, latency, and accuracy. The table shows that an analog SNN uses less than half the energy and has much lower latency compared to the other configurations. These results support the hypothesis set out above: analog-mixed-signal

SNN configurations can achieve ultra-low power consumption, fast response time, and comparable accuracy for intelligent-device applications. Therefore, for the devices running on a limited battery supply and for continuous 24/7 monitoring, such an improvement is highly desirable.

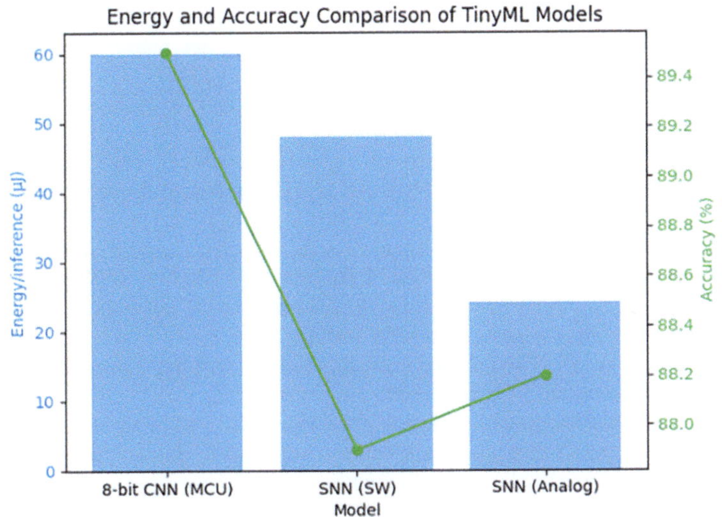

Fig. 2. Energy and accuracy.

References

1. Author, F., et al.: From tiny machine learning to tiny deep learning: a survey. ACM Comput. Surv. (2025). In press
2. Author, F., et al.: Standardized latency and energy measurement methodology for Tinyml. J. Low Power Electr. Appl. (2025). Preprint
3. Banbury, C.R., et al.: Benchmarking Tinyml systems: challenges and directions. arXiv preprint arXiv:2103.03148 (2021)
4. Banbury, C., Reddi, V.J., et al.: Mlperf tiny benchmark. arXiv preprint arXiv:2106.07597 (2021)
5. Gopalakrishnan, S., Chatterjee, A.: Neuromorphic computing: mixed-signal neurons for ultra-low-power inference. IEEE J. (2018)
6. Gopalakrishnan, S., Chatterjee, A.: Neuromorphic computing: mixed-signal neurons for ultra-low-power inference. IEEE J. (2018). https://arxiv.org/ftp/arxiv/papers/1806/1806.05141.pdf
7. Heydari, S., et al.: Tiny machine learning and on-device inference: a survey of shared challenges and technologies. Sensors **25**(5), 1870 (2025)
8. Heydari, S.E.A.: Tiny machine learning and on-device inference: a survey of shared challenges and technologies. Sensors **25**(5), 1870 (2025)

9. Indiveri, G., Liu, S.C.: Memory and information processing in neuromorphic systems. Proc. IEEE **103**(8), 1379–1397 (2015)
10. Kösters, U., et al.: Energy-efficient low-latency spiking neural networks using analog in-memory computing. APL Mach. Learn. **1**(2), 026101 (2023)
11. Lin, J., et al.: Mcunet: Tiny deep learning on IoT devices. In: NeurIPS (2020)
12. Maass, W.: Networks of spiking neurons: the third generation of neural network models. Neural Netw. **10**(9), 1659–1671 (1997)
13. Ostrau, C., et al.: Benchmarking neuromorphic systems: energy, latency, and accuracy across platforms. Front. Neurosci. **16**, 871990 (2022)
14. Roy, K., et al.: Towards spike-based machine intelligence with neuromorphic computing. Nature **575**, 607–617 (2019)
15. Shafique, M., et al.: Tinyml: current progress, research challenges, and future roadmap. In: Proceedings of the 58th ACM/IEEE Design Automation Conference (DAC) (2021)
16. Song, J., et al.: Time-domain neuron circuits for ultra-low-latency spiking neural networks. IEEE Trans. Circuits Syst. I Regul. Pap. **69**(11), 4600–4613 (2022)
17. Sorbaro, M., et al.: Optimizing energy in spiking neural networks with spike-based computation. Front. Neurosci. **14**, 658 (2020)
18. Stewart, R.D., et al.: Evaluating encoding and decoding approaches for spiking neuromorphic systems. Neuromorphic Eng. (2022)
19. Stewart, R.D.e.a.: Evaluating encoding and decoding approaches for spiking neuromorphic systems. Neuromorphic Engineering (2022). https://neuromorphic.eecs.utk.edu/publications/2022-07-28-evaluating-encoding-and-decoding-approaches-for-spiking-neuromorphic-systems/
20. Tan, M., Le, Q.V.: Efficientnet: Rethinking model scaling for convolutional neural networks. Proceedings of the 36th International Conference on Machine Learning, pp. 6105–6114 (2019)
21. Tavanaei, A., Ghodrati, M., Kheradpisheh, S.R., Masquelier, T., Maida, A.: Deep learning with spiking neurons: opportunities and challenges. Front. Neurosci. **12**, 774 (2018)
22. Tavanaei, A., et al.: Deep learning in spiking neural networks. Neural Networks (2019)
23. Wang, Z., et al.: Accurate and efficient ANN-to-SNN conversion through threshold balancing. Proceedings of the International Joint Conference on Artificial Intelligence (IJCAI), pp. 567–574 (2023)
24. Warden, P., Situnayake, D.: Tinyml: A survey of machine learning in embedded systems. In: Embedded Systems Conference (2019)
25. Wu, Y., et al.: Efficient spiking neural network conversion with explicit current control. Front. Neurosci. **16**, 879457 (2022)
26. Xiao, L., et al.: Co-design in Tinyml: addressing microcontroller limits. IEEE Circ. Syst. Mag. 10–20 (2023)
27. Zhang, K., et al.: Characterizing the performance and energy of Tinyml on microcontrollers. In: IEEE International Conference on Computer Design (ICCD), pp. 191–198 (2022)

Open Access

Open Access This chapter is licensed under the terms of the Creative Commons Attribution-NonCommercial-NoDerivatives 4.0 International License (http:// creativecommons.org/licenses/by-nc-nd/4.0/), which permits any noncommercial use, sharing, distribution and reproduction in any medium or format, as long as you give appropriate credit to the original author(s) and the source, provide a link to the Creative Commons license and indicate if you modified the licensed material. You do not have permission under this license to share adapted material derived from this chapter or parts of it.

The images or other third party material in this chapter are included in the chapter's Creative Commons license, unless indicated otherwise in a credit line to the material. If material is not included in the chapter's Creative Commons license and your intended use is not permitted by statutory regulation or exceeds the permitted use, you will need to obtain permission directly from the copyright holder.

Polarity-Weighted Semantic Expansion with Multinomial Naive Bayes Ensemble for Robust Fake News Detection in Social Media Platforms

G. Revathi[✉] and G. K. Arun

Arignar Anna Government Arts College, Villupuram, Tamil Nadu, India
revathigovind1987@gmail.com, arunnura2370@gmail.com

Abstract. Social media is full of misinformation that tugs at the heartstrings but doesn't say anything definite. This study introduces the Polarity-Weighted Semantic Expansion coupled with a Multinomial Naive Bayes Ensemble (PWSE-MNB-E) as a resilient approach for fake-content identification. Core components sentiment-sensitive lexical extraction, semantic condensation, polar-transfer, and frequency-weighting are harmonised within a Naive Bayes framework, supplemented by redundancy curtailment and ensemble moderation to enhance predictive accuracy. The Kaggle Fake-Content repository has been subjected to multiple empirical tests and it is evaluated that PWSE-MNB-E attains an accuracy of 63.36% and an F-Measure of 64.35%, advancing over the MMFND and SARD-FN benchmark models. The F Measure Index and Matthews Correlation Coefficient Indexes also register improvement indexes at 64.48 and 27.21 respectively indicating improvement in reliability in the case of highly unbalanced class distributions. The experiment strongly supports the architecture's ability to capture subtle variation in affect-phraseing and thus offers an effective, low-cost means of counteracting deliberate disinformation.

Keywords: Fake News Detection · Sentiment Analysis · Semantic Expansion · Multinomial Naive Bayes · Ensemble Classification · Misinformation Identification

1 Introduction

The increase in multimodal fake news poses a significant threat to digital media communication since it is known that fake news propagates quickly and accompanies various formats, like texts and images and videos, and alters the opinion of the public and distorts social awareness [1]. This has been largely fuelled by the accessibility and the virality of social media platforms which have amplified the spread of such gun weaponization narratives. These are invented material which is usually designed to appeal on an emotional level, to entrench some prejudices, or sometimes to push tracts of ideology. Specifically, the news stories that are socially constructed and have the capacity of remaining plausible based on such aspects as the use of stylised headlines, sensational language, and

© The Author(s) 2026

J. C. Bansal et al. (Eds.): SCIS 2025, LNNS 1929, pp. 128–141, 2026.
https://doi.org/10.1007/978-3-032-22911-3_10

source ambiguity, have even proved to be more difficult to notice in comparison with the conventional verification techniques [2].

1. The trend of fake news propagation has impacted the society at large by affecting the outcomes of democratic processes, health behaviour of citizens and international perceptions. Opinion-manipulation is invited regularly by users of digital media through content that is made to appear factual like news, creating partisanism and misguidedness [3]. It has become imperative, as the same has come up because of the necessity of scalable and data-based methods of finding and countering such content. With evolving approaches in misinformation, content-based systems of detection require revision to deal with the complexity, ambiguity, and emotional deception that are built into the textual structures, and cross-modal ties [4].

2. Deep learning and machine learning have provided a paramount contribution to this challenge by providing the idea of automated detection systems which can acquire discriminative features by labelled examples. Linguistic signals, content propagation and user behaviour have been analysed using traditional classifiers, neural architecture based on sequences, as well as graph-based models [5]. Such methods utilize natural language comprehension and situation modeling in separating the trustworthy and false stories. Nevertheless, it remains rather difficult to find subtle manipulative emotional content, vocabulary drift, and changes in topics employed by the creators of fake news in order to avoid detection [6].

3. Multinomial Naive Bayes algorithm as a simple and easy-to-interpret method, although is no longer used in text classification, may become a useful tool when the computational efficiency is essential. It is relatively reliable since it can model word frequencies and come up with probabilistic class scores. Polarity-Weighted Semantic Expansion (PWSE), as a technique used to make a given detector more competent in terms of detecting fake news, adds a sentiment-level to increase the impact of the information about the emotions and semantics [7]. PWSE contributes to the lexical space filling the polarity-charged token through a semantically related term, weighted with respect to affect intensity. This multisided improvement provides a language-based route structure to be more able to isolate the feeling-based representation that commonly lurks in fake news [8]. Bio-inspired optimizations are utilized to achieve the desired cumulative outcomes along with improving overall results.

1.1 Problem Statement

The swift spread of fake news on the digital platforms has made the challenge of spreading reliable information even difficult. The misinformation through text tends to incorporate sentiment manipulation and semantic ambiguity in order to confuse the reader. These subtle tendencies are overlooked in the traditional models, particular instances where the lexical deviations conceal emotional bias. The majority of the detection systems do not pay attention to the semantic proximity and sentiment intensity of alternative terms, they become less sensitive to the deceptive content. A model that combines both sentiment awareness and lexical generalisation becomes critical in a model with high on interpretability. To respond to these limitations, the approach is necessary that would be able to grasp not only the emotional polarity but also the context-based term replacement in fake news.

1.2 Motivation and Objective

Fake news is particularly present in emotionally charged language that aims to manipulate the perception, found in the short form of content that propagates via the digital channel. Such manipulation does not merely depend on deceptive information but also on the use of minor language work such as polarised sentiment, suggestive tone, and variation through use of synonyms that are not captured in conventional detection models. The sentiment strength and semantic proximity of these advertising constructs remain uninterpretable by standard classifiers, and thus lead to poor sensitivity in the level of detection. It has been driven by this challenge that this work aims to provide an improved probabilistic framework by based the Multinomial Naive Bayes model on PWSE. The proposal to be examined in this chapter will center on the identification of sentiment-laden terms, semantically bounding them with domain-specific semantic neighbors, and adjusting the impacts with the use of polarity indicators. With such integration, the model would lexically and contextually discern the manipulative intention and variety of fakes, as well as the diverse range of fakes and misinformation circulated in social networks.

2 Literature Review

The review intentionally foregrounds studies published in 2024 in order to portray the most recent advancements in the field of misinformation detection; in so doing, the survey strengthens its empirical foundation, guarantees topicality, and ensures practical applicability to contemporary scholarly and operational discourse.

"Layered Voting Block" [9] used TF-IDF and Bag-of-Words features with logistic regression, decision trees, and SVM. Model outputs were merged through a weighted voting mechanism. Noise reduction and bootstrapped sampling enhanced diversity. Classifier blending reduced overfitting, delivering stable fake news detection by balancing lexical evidence with ensemble diversity. "SemNet Explorer" [10] built a semantic-driven architecture combining convolutional extractors, graph embeddings, and BiLSTM. Word-level dependencies and paragraph structure were encoded. Semantic misalignments were flagged via attention focus. The "PolCred Analyzer" [11] incorporates metadata, linguistic structure, and SHAP/LIME explanation in a transparent framework. Narratives processed through attention LSTM and metadata guided interpretability. Credibility scoring captured article distortion. According to [12], Hy-brid Squeeze Summary merged both extractive and abstractive summaries, which generated an LSTM decoder and then further compressed the result using statistical filters. Knowledge-transferred embeddings enhanced classification. Subword handling resolved dialect variance. The strategy focused on compact inputs with rich semantics for low-resource misinformation detection problems. The researchers in [13] evaluated GloVe, FastText, and Word2Vec through deep and classical classifiers using the "Embed-Classifier Matrix". We ran tests for coherence using embeddings in paired architectures. Concatenated vectors improved context spread.

The "DANES Context Stack" [14] combined multiple encoders to capture semantic and social cues by integrating the BiLSTM (Bidirectional Long Short-Term Memory), feedforward encoder, and a social graph encoder (SGE). Voting coordination was used to aggregate predictions of different sub-networks. User response normalization added external evi-dence. The "GRU Claim Tracker" [15] used stacked GRUs to encode long-range dependencies. Gating preserved transition cues across narrative segments. The feature vectors were shaped by tone shifts and contradiction indicators. Dense layers finalized outputs. "Modal Dominance Fusion" [16] handled content like text, images, metadata, and propagation separately. Fusion gates selected prominent features through inter-modal attention. Irrelevant cues were filtered. The classifier adapted to modality shifts, allowing flexible detection of synthetic content across various media types. TextVision Synergy [17] combined graph plus text representation with transformer plus image analyses.Co-occurrence graphs and visual patterns aligned through cross-view attention. Joint embedding space supported modality fusion. Hierarchical architecture captured mutual reinforcement between textual and visual elements in deceptive media content. "SARD-FN" [18] used CLIP with contrastive learning to align image-text pairs for fake news spotting. A projection head refined embeddings, optimizing similarity for real pairs and penalizing mismatched ones using cosine divergence. A fusion module combined multimodal cues before classification. "MMFND" [19] employed Siamese encoders for coherence scoring across modalities. A discriminator identified fake alignments while a generator simulated manipulative patterns. Attention maps emphasized deceptive content. Final classification relied on joint adversarial and similarity loss.

Despite strong progress in multimodal and deep learning approaches, limited research has directly integrated polarity-sensitive semantic expansion with probabilistic classifiers, which motivates the present PWSE-MNB-E framework.

3 Polarity-Weighted Semantic Expansion with Multinomial Naive Bayes Ensemble (PWSE-MNB-E)

PWSE-MNB-E refers to a hybrid classification model that enriches lexical features using sentiment-guided expansion and combines probabilistic inference with ensemble learning for effective fake news detection.

3.1 Sentiment-Driven Term Extraction

Sentiment-Driven Term Extraction is meant to perform a targeted extraction of sentiment-bearing words in news material, which can be early warning signs of emotional or agenda-pushing motion, as is seen in created narratives. In the architecture of PWSE-MNB-E, the step assumes a prominent position as the input is prepared ready to be augmented with polarity-aware enhancing. Tokens that are rich in emotion that usually appear in fake news, especially in news streams on social media are preoccupying with the opinion of people with inflated or inaccurate overtones. Such tokens are therefore isolated to provide a simple filter on which subsequent expansion and weighting processes can be carried out more accurately.

Let $D = \{d_1, d_2, \ldots, d_n\}$ be the document set and V the vocabulary extracted after initial tokenization. Each document d_i undergoes lexicon-based sentiment probing using a polarity-scored lexicon L, which maps a term $t_k \in V$ to a polarity weight $\psi_k \in [-1,1]$. A sentiment selection function $S(d_i)$ maps a document to a polarity token subset as

$$S(d_i) = \{t_k \in d_i \mid |\psi_k| \geq \delta\} \tag{1}$$

where $S(d_i)$ denotes the polarity-filtered term set from document d_i, and δ acts as a semantic gating constant. A higher δ filters out lexically neutral or weakly polar terms, improving the focus on emotionally salient signals. The threshold δ is optimized using validation feedback from early-stage model iterations.

To further quantify the emotional skewness embedded within a document, a document-level polarity distribution index $PDI(d_i)$ is calculated as

$$PDI(d_i) = \frac{1}{|S(d_i)|} \sum_{t_k \in S(d_i)} \psi_k \cdot log(1 + f_k) \tag{2}$$

where $PDI(d_i)$ indicates the sentiment dominance score of document d_i, and f_k refers to the term frequency of token t_k. The logarithmic scaling component mitigates overemphasis from high-frequency sentiment words that may occur due to repetition rather than intensity. This scalar index affects the polarity calibration strength applied in later steps of the PWSE-MNB-E model, particularly during expansion and weighting.

3.2 Domain-Constrained Semantic Expansion

Domain-constrained semantic expansion is critical for transforming sentiment-bearing terms into a richer, semantically expressive space aligned with the discourse of fake news. In PWSE-MNB-E, this step extends each extracted polarity-laden token into a cluster of semantically related terms, carefully filtered based on domain specificity. Such expansions are essential in the context of sentiment-laden tweets, where manipulative content may exploit lexical variation while maintaining deceptive intent. Semantic augmentation ensures robustness against lexical sparsity and vocabulary mismatch in short, opinionated news posts.

Let $\Theta = \{t_1, t_2, \ldots, t_m\}$ be the set of sentiment-bearing tokens extracted from Step 1. For each token $t_i \in \Theta$, an embedding-based similarity search retrieves top-k candidate neighbors $N_{t_i} = \{w_1, w_2, \ldots, w_k\}$ based on cosine proximity

$$sim(t_i, w_j) = \frac{v(t_i) \cdot v(w_j)}{\|v(t_i)\| \cdot \|v(w_j)\|} \tag{3}$$

where $v(t_i)$ and $v(w_j)$ denote the word embeddings of t_i and its semantic neighbor w_j, respectively. A candidate is retained only if $sim(t_i, w_j) \geq \gamma$, with γ being a tuned similarity threshold that ensures semantic proximity relevant to deceptive sentiment patterns.

To further filter these neighbors by contextual relevance, each candidate term undergoes subject-alignment evaluation using the document's metadata. A subject-matching score $\rho(w_j)$ is computed as

$$\rho(w_j) = \frac{TFIDF(w_j) \cdot TFIDF(S_d)}{\|TFIDF(w_j)\| \cdot \|TFIDF(S_d)\|} \tag{4}$$

where $TFIDF(S_d)$ represents the vectorized subject line associated with the news content. Terms with $\rho(w_j) \geq \eta$, where η is the subject relevance threshold, are maintained to make sure that expansion is domain-congruent with the article.

The set of final validated semantic extensions of each seed token is provided by

$$\varepsilon_{t_i} = \left\{ w_j \in N_{t_i} | sim(t_i, w_j) \geq \gamma \wedge \rho(w_j) \geq \eta \right\} \tag{5}$$

where ε_{t_i} denotes the domain-constrained expanded token cluster for t_i. The conjunctive filtering reasoning imposes semantic consistency as well as subject compatibility. The inclusion of embedding-level similarity together with metadata-driven constraints assures domain precision in PWSE-MNB-E and avoids semantic drift that often occurs when the expansion is uncontrolled.

3.3 Polarity Propagation to Expanded Terms

Propagation to Expanded Terms helps to ensure that the emotional weight possessed by the original tokens is not lost at the token's semantic neighborhood. In the PWSE-MNB-E approach, the proportionality mechanism ensures that each domain-consistent expansion created in the previous step carries emotional charge proportional to the sentiment load of its point of origin. Misleading news especially those that are circulated on the social media tend to use synonym variations to disguise emotional sentiments and thus leave behind very deceptive emotional impressions. Such disguise is undone in this step by letting expanded terms inherit calibrated polarity and then sentiment continuity is allowed to blanket the feature space.

For each seed token t_i with polarity weight ψ_i, all its domain-filtered semantic expansions $w_j \in \varepsilon_{t_i}$ are ascribed values of polarity by a propagation rule that is given by

$$\phi(w_j) = \psi_i \cdot sim(t_i, w_j) \tag{6}$$

where $\phi(w_j)$ measures the propagated polarity of term w_j, ψ_i is the original sentiment score and t_i, and $sim(t_i, w_j)$ the cosine similarity already calculated during semantic expansion. In this equation, the emotional influence decays slowly in the case of semantically distant expansions, whereas, it is strong in the case of close matches.

In order to normalise the polarity effect among various seed terms in common expanded token w^*, an aggregate polarity index is estimated as a result of computing the following index along aggregate index.

$$\Phi(w^*) = \frac{1}{|T_{w^*}|} \sum_{t_i \in T_{w^*}} \phi_{t_i}(w^*) \tag{7}$$

where $\Phi(w^*)$ represents the average propagated polarity of expanded token w^*, and T_{w^*} is the set of all seed terms t_i from which w^* was derived. The score $\phi_{t_i}(w^*)$ is the polarity score of the specific seed term to the common expanded token. This formulation puts into consideration the impact of several seeds and maintains directional sentiment cues.

3.4 Weighted Frequency Transformation

The Weighted Frequency Transformation is one of the most important steps in PWSE-MNB-E pipeline, since it allows adjusting the values of the token frequencies by directly incorporating sentiment strength into the statistical core of the classifier. Particular terms may find deceptive force not through a one-dimensional repetition but through the affective force that they draw in the context of fake news disseminated on social platforms. The given transformation phase will guarantee that all terms that are either original or semantically extended show not only the number of them but also the emotional values that they embody.

Let $f\left(w_j, d_i\right)$ denote the raw frequency of a token w_j in a document d_i, and let $\phi\left(w_j\right)$ is the sentiment polarity score of token based on the propagation mechanism in Step 3. The polarity-weighted frequency is, given by $\omega\left(w_j, d_i\right)$.

$$\omega\left(w_j, d_i\right) = f\left(w_j, d_i\right) \cdot \left(1 + \lambda \cdot \left|\phi\left(w_j\right)\right|\right) \tag{8}$$

where λ is a polarity amplification constant that was set via cross validation to regulate the influence of sentiment upon the scaling of frequencies. This modification makes it possible that emotionally charged words, either synthetic or natural are accredited more prominence in classification as not to disfigure the neutral a textual framework

In order to alleviate the domination effect of high-frequency polar terms, a logarithmic smoothing function is used to smooth out the resulting contribution score as balanced contribution score $\xi\left(w_j, d_i\right)$

$$\xi\left(w_j, d_i\right) = log\left(1 + \omega\left(w_j, d_i\right)\right) \tag{9}$$

Where $\xi\left(w_j, d_i\right)$ refers to the smoothed token weight of its sentiment. Such a refined frequency vector will form the new feature set being passed into the Naive Bayes estimator, closing the gap between sentiment semantics and statistical learning.

3.5 Redundancy Suppression via Semantic Merging

The inflation of semantically identical or near-duplicate terms that can be caused by sentiment-guided expansion is mitigated in PWSE-MNB-E architecture with the use of Redundancy Suppression via Semantic Merging. In the absence of semantic consolidation, the models of fake news detection face a danger of overfitting towards lexical variants of words which represent essentially the same semantics. Such issue is being especially exacerbated on short and emotionally loaded news posts on social media, as the semantic diversity can be quite low, and the reiteration in wording is frequent. With successful merging, there must be compactness in feature space, weight distribution is balanced and classes become more distinct.

Let w_a and w_b be any two polarity-expanded tokens. A semantic equivalence check is performed using vector cosine similarity. A merging condition $M\left(w_a, w_b\right)$ is defined as

$$\left(w_a, w_b\right) = \begin{cases} 1, & if\ \frac{v(w_a) \cdot v(w_b)}{\|v(w_a)\| \|v(w_b)\|} \geq \theta \\ 0, & otherwise \end{cases} \tag{10}$$

where $M\left(w_a, w_b\right) = 1$ denotes a merge trigger based on cosine similarity exceeding a threshold θ, tuned within the range [0.85, 0.95].

For all token pairs meeting the merge condition, their weighted frequencies $\omega(w_a, d_i)$ and $\omega(w_b, d_i)$ from Step 4, the terms are combined into a unified term representation as

$$\Omega(w_a, w_b) = \omega(w_a, d_i) + \omega(w_b, d_i) \tag{11}$$

Where $\Omega(w_a, w_b)$ replaces both w_a and w_b in the feature vector, the lower-frequency term is discarded to avoid duplication.

To maintain polarity integrity, the merged polarity score $\Psi(w_a, w_b)$ is calculated as

$$\Psi(w_a, w_b) = \frac{\phi(w_a) \cdot \omega(w_a, d_i) + \phi(w_b) \cdot \omega(w_b, d_i)}{\omega(w_a, d_i) + \omega(w_b, d_i)} \tag{12}$$

where $\Psi(w_a, w_b)$ represents the polarity-weighted average, preserving sentiment influence after term fusion. This redundancy-aware semantic reduction ensures that PWSE-MNB-E maintains a concise yet sentiment-strong representation of fake news texts, minimising noise without sacrificing discriminatory power.

3.6 Polarity-Calibrated Naïve Bayes Estimation

Polarity-Calibrated Naïve Bayes Estimation appends a sentiment-sensitive probabilistic classification layer to PWSE-MNB-E by recalibrating the likelihood computations. Rather than implementing a frequency-based model of computing the likelihood of an event by assuming that all terms are independent of each other and that all terms equally contribute to the captured model, this approach, the polarity recalibrated approach, focuses on the emotionally salient words and phrases captured during an expansion and merging the weighted model. Within the context of the misinformation propagated through social medial platforms, news reports with emotionally charged content tend to bias the users' perceptions by reframing perceptions with an emphasis on emotionally charged words. This step, the emotionally charged content and this step in the Naive Bayes model, is what is referred to in this instance.

For each class $c \in \{fake, real\}$, An estimate of the polarity-weighted conditional likelihood of observing a word w_j is

$$P(w_j|c) = \frac{\xi(w_j, c) + \alpha}{\sum_{w \in V_c} \xi(w_j, c) + \alpha \cdot |V_c|} \tag{13}$$

where $\xi(w_j, c)$ is the polarity-smoothed frequency derived after transformation and redundancy suppression, V_c is the class-specific vocabulary, and α is the smoothing constant for unseen terms. This formulation adjusts the likelihood landscape in favour of sentiment-relevant vocabulary.

The log-likelihood for a document d_i belonging to class c is computed using additive scoring.

$$L(d_i, c) = logP(c) + \sum_{w_j \in d_i} \xi(w_j, d_i) \cdot logP(w_j|c) \tag{14}$$

where $L(d_i, c)$ denotes the document-wise class log-probability, $P(c)$ is the prior probability of class c, and $\xi(w_j, d_i)$ represents the sentiment-weighted occurrence of term w_j in document d_i. This scoring function introduces polarity-aware emphasis to the standard log-likelihood structure.

To calibrate the final classification decision under polarity guidance, the posterior estimation integrates a sentiment modulation factor β to yield

$$\hat{c} = \arg\max_c [L(d_i, c) \cdot (1 + \beta \cdot mean(\Phi(d_i)))] \tag{15}$$

where $\hat{c}$ denotes the predicted class label, and $(\Phi(d_i))$ is the average propagated polarity score of all active terms in document d_i. The factor β amplifies classification influence from sentiment-dense content, ensuring that PWSE-MNB-E reflects emotionally manipulative patterns prevalent in fabricated news narratives.

3.7 Ensemble-Based Decision Scaling

Ensemble-Based Decision Scaling constitutes the final optimisation layer within PWSE-MNB-E, designed to stabilise and refine the classification output by integrating probabilistic cues from multiple discriminative sources. While the polarity-calibrated Naïve Bayes estimator captures sentiment-weighted lexical cues, auxiliary classifiers trained on raw metadata and structure-specific embeddings contribute contextual grounding. In the landscape of social media-driven fake news, sentiment alone may misfire due to sarcasm, domain ambiguity, or impersonated credibility. Ensemble scaling mitigates such imbalance by assigning calibrated importance to each predictive stream.

Let $L_{NB}(d_i, c)$ denote the log-likelihood score for class c from the polarity-aware Naïve Bayes model. Independently, let a secondary metadata-guided model produce a class confidence score $S_{LR}(d_i, c)$ derived from logistic regression or shallow decision models. The ensemble-scaled probability for class c is defined as

$$(d_i, c) = \lambda \cdot \sigma(L_{NB}(d_i, c)) + (1 - \lambda) \cdot S_{LR}(d_i, c) \tag{16}$$

where $P_{ens}(d_i, c)$ represents the fused class probability, σ is the softmax normalisation over the Naïve Bayes log-likelihoods, and $\lambda \in [0,1]$ is the ensemble weighting factor is empirically tuned via cross-validation.

To enhance polarity alignment across classifiers, a polarity-resonance penalty $\Delta_c(d_i)$ is incorporated into the final scoring function.

$$\Delta_c(d_i) = |S_{LR}(d_i, c) - \sigma(L_{NB}(d_i, c))| \cdot \mu(d_i) \tag{17}$$

Where $\Delta_c(d_i)$ quantifies the disagreement between classifiers scaled by document-level polarity intensity $\mu(d_i)$, derived from mean sentiment weights. Penalising significant polarity divergences anchors the ensemble to emotionally relevant predictions.

The final decision rule selects the class with the highest scaled probability corrected for polarity divergence

$$\hat{c} = \arg\max_c [P_{ens}(d_i, c) - \Delta_c(d_i)] \tag{18}$$

where $\hat{c}$ indicates the optimised label output. This polarity-sensitive ensembling reinforces the PWSE-MNB-E model's ability to detect nuanced deception patterns across text, metadata, and sentiment perspectives in socially propagated news content.

This integration connects interpretable probabilistic models with sentiment-informed semantic enrichment so as to balance explanatory clarity and operational efficacy.

4   Dataset Details and Outcomes

The Kaggle Fake News Detection benchmark corpus encompasses a total of 44,898 documents, each labeled as either deceptive (fake) or authentic (real). A comprehensive description of the dataset is grounded in its textual and metadata constituents, cataloguing for each instance the headline, the complete body of the article, the author, the subject category, and the timestamp of publication. These attributes furnish rich contextual layers for both contextual analysis and polarity-driven semantic modeling. The corpus accommodates binary classification tasks as well as tasks sensitive to dynamic sentiment shifts, thereby enabling a breadth of inquiries focused on semantic amplification, lexical dimorphism, and the affective dimensions of linguistic exploitation observable in fabricated journalistic materials. Its structured format is conducive to seamless coupling with both classical and deep-learning methodologies directed toward the detection of misinformation in contemporary digital communication waterfalls. The baseline architectures SARD-FN [18] and MMFND, together with the specific dataset and experimental configuration delineated in this section, serve as the comparative benchmark upon which all subsequent results are evaluated. The experimental framework operationalizes the Kaggle Fake News corpus, employing the SARD-FN [18] and MMFND [19] benchmarks as comparative standards. Training protocols applied an 80/20 stratified partition of the data, complemented by the conventional smoothing procedure of the Naive Bayes classifier. Quantitative assessment entailed computation of accuracy, F-measure, the Fowlkes-Mallows Index (FMI), Matthews Correlation Coefficient (MCC), precision, and recall metrics.

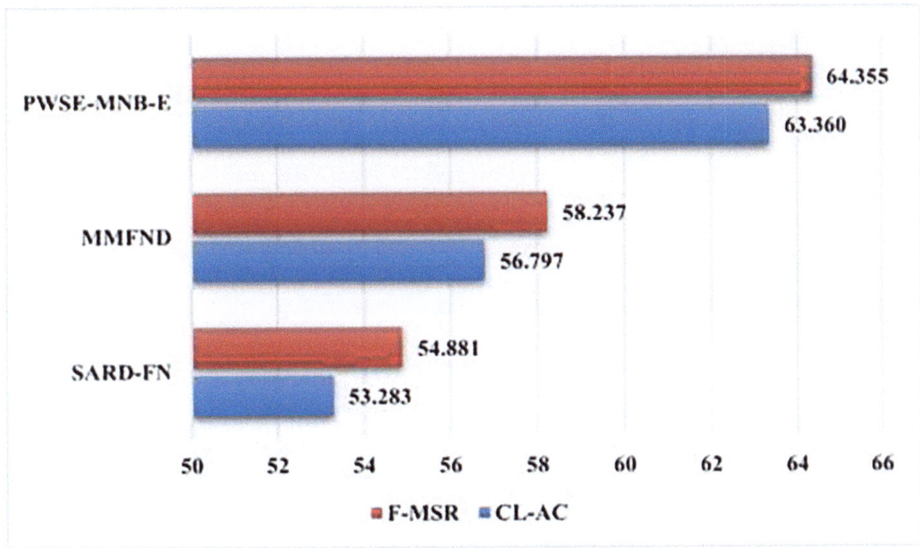

Fig. 1. CL-AC and F-MSR

Results and discussion centers on comparing the performance of the classification ability of the proposed model to that of the available baselines with standard performance measures. Classification Accuracy (CL-AC) incurs the ration of those instances that were correctly classified to the amount of instances, and the F-Measure (F-MSR) which is the harmonic mean of precisions stratified by the recall score, is the correspondence of good positive identification and falsely negative downgrade. In terms of Classification Accuracy, PWSE-MNB-E achieved 63.360%, outperforming MMFND at 56.797% and SARD-FN at 53.283%. Regarding the F-Measure, PWSE-MNB-E recorded 64.355, showing a significant improvement over MMFND with 58.237 and SARD-FN with 54.881. These results demonstrate superior consistency in correctly classifying both fake and real news using polarity-weighted expansion which is pictorially illustrated in Fig. 1.

Figure 2 shows the evaluation based on Fowlkes–Mallows Index (FMI) and Matthews Correlation Coefficient (MCC) highlights the effectiveness of the classification models in handling fake news detection tasks. FMI reflects the geometric mean of precision and recall, while MCC indicates the overall quality of binary classifications, especially under class imbalance. PWSE-MNB-E attained an FMI of 64.488 and MCC of 27.215, which are considerably higher than MMFND (FMI: 58.445, MCC: 14.363) and SARD-FN (FMI: 54.938, MCC: 6.686). The higher scores from PWSE-MNB-E confirm stronger agreement between predicted and actual classes, validating its robustness in identifying sentiment-driven misinformation patterns.

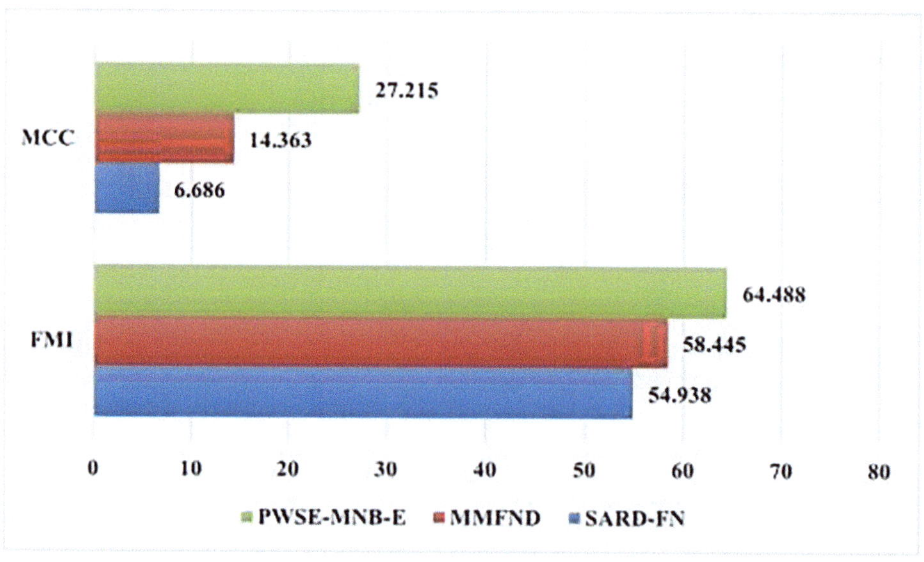

Fig. 2. FMI and MCC

The quality of the predictions of fake news classification is determined by the Precision (PREC) and Recall (RCLL) values. Precision calculates how many of the fake news are properly labeled as such as compared to the number of fake news predicted, and Recall measures the likelihood of capturing all real fake cases. PWSE-MNB-E returned the best values of Precision 60.481% and Recall 68.759% showing better repeatability in detecting fake content. Precision and Recall using MMFND was 53.708% and 63.600%, respectively, and SARD-FN, 52.490% and 57.499%, respectively. These findings show that: PWSE-MNB-E works better in maintaining a balance between accuracy and relevance in detection of sentiment-filled fake news in online platforms as shown pictorially in Fig. 3.

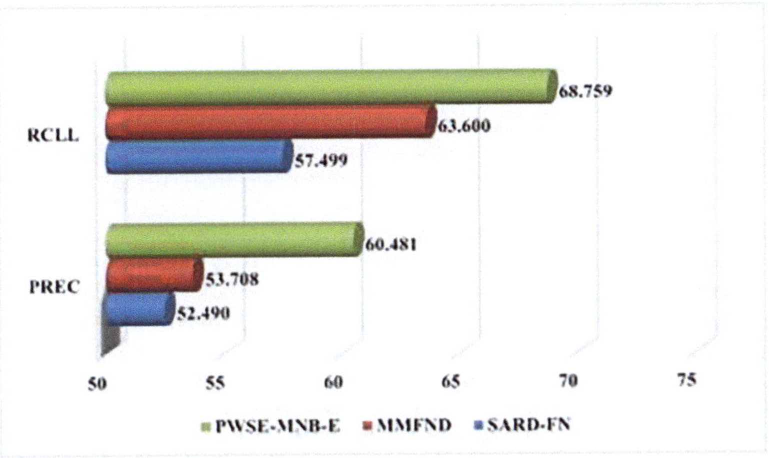

Fig. 3. PREC and RCLL

5 Conclusion

PWSE-MNB-E suggested framework creates a sentiment-accentuating procedure of handling lexical ambiguity, at the presence of emotional manipulation in fake news that is being circulated through digital media. By integrating polarity-weighted semantic expansion with probabilistic estimation, the model strengthens contextual awareness without increasing computational complexity. Domain-constrained expansion, polarity propagation, and frequency transformation collectively enhance feature richness while maintaining model interpretability. Semantic merging suppresses redundancy, and ensemble scaling balances structural and emotional cues. This design aligns with the linguistic and psychological nature of misinformation, especially in socially shared content. The method offers a linguistically grounded and computationally efficient framework capable of capturing sentiment-driven deception patterns. The complete structure supports broader applicability in misinformation analysis across multiple communication layers. Future work will extend PWSE-MNB-E towards multimodal fusion across text, image, and video, as well as adaptation for under-resourced languages to generalise its impact.

References

1. Gemeda yigezu, M., Mersha, M.A., Bade, G.Y., Kalita, J., Kolesnikova, O., Gelbukh, A.: Ethio-fake: cutting-edge approaches to combat fake news in under-resourced languages using explainable AI,. Procedia Comput. Sci. **244**, 133–142 (2024). https://doi.org/10.1016/j.procs.2024.10.186

2. Fang, X., et al.: NSEP: early fake news detection via news semantic environment perception. Inf. Process. Manag. **61**(2), 103594 (2024). https://doi.org/10.1016/j.ipm.2023.103594

3. Dinu, L., Fusu, E.C., Gifu, D.: Veracity analysis of romanian fake news. Procedia Comput. Sci. **225**, 3303–3312 (2023). https://doi.org/10.1016/j.procs.2023.10.324

4. Aman, M.: Large language model based fake news detection. Procedia Comput. Sci. **231**, 740–745 (2024). https://doi.org/10.1016/j.procs.2023.12.144

5. Fakieh, B., AL-Malaise AL-Ghamdi, A.S., Saleem, F., Ragab, M.: Optimal machine learning driven sentiment analysis on COVID-19 twitter data. Comput. Mater. Continua **75**(1), 81–97 (2023). https://doi.org/10.32604/cmc.2023.033406

6. Zhang, M., Chai, J., Cao, J., Ji, J., Yi, T.: Aspect-level sentiment analysis based on deep learning. Comput. Mater. Continua **78**(3), 3743–3762 (2024). https://doi.org/10.32604/cmc.2024.048486

7. Nadeem, M.I., et al.: SSM: Stylometric and semantic similarity oriented multimodal fake news detection. J. King Saud Univ.- Comput. Inform. Sci. **35**(5), 101559 (2023). https://doi.org/10.1016/j.jksuci.2023.101559

8. Mohawesh, R., Liu, X., Arini, H.M., Wu, Y., Yin, H.: Semantic graph based topic modelling framework for multilingual fake news detection. AI Open **4**, 33–41 (2023). https://doi.org/10.1016/j.aiopen.2023.08.004

9. Ilyas, M.A. Rehman, A., Abbas, A., Kim, D., Naseem, M.T., Allah, N.M.: Fake news detection on social media using ensemble methods. Comput. Mater. Continua **81**(3), 4525–4549 (2024). https://doi.org/10.32604/cmc.2024.056291

10. Alghamdi, J., Lin, Y., Luo, S.: Unveiling the hidden patterns: a novel semantic deep learning approach to fake news detection on social media. Eng. Appl. Artif. Intell. **137**, 109240 (2024). https://doi.org/10.1016/j.engappai.2024.109240

11. Kozik, R., et al.: Towards explainable fake news detection and automated content credibility assessment: polish internet and digital media use-case. Neurocomputing **608**, 128450 (2024). https://doi.org/10.1016/j.neucom.2024.128450

12. Alghamdi, J., Lin, Y., Luo, S.: Fake news detection in low-resource languages: a novel hybrid summarization approach. Knowl Based Syst **296**, 111884 (2024). https://doi.org/10.1016/j.knosys.2024.111884

13. Hauschild, J., Eskridge, K.: Word embedding and classification methods and their effects on fake news detection. Mach. Learn. Appl. **17**, 100566 (2024). https://doi.org/10.1016/j.mlwa.2024.100566

14. Truică, C.-O., Apostol, E.-S., Karras, P.: DANES: deep neural network ensemble architecture for social and textual context-aware fake news detection. Knowl. Based Syst. **294**, 111715 (2024). https://doi.org/10.1016/j.knosys.2024.111715

15. Elfaik, H., Nfaoui, E.H.: Automatic detection of fake news using gated recurrent unit deep model. Procedia Comput. Sci. **233**, 474–480 (2024). https://doi.org/10.1016/j.procs.2024.03.237

16. Fu, L., Peng, H., Ma, C., Liu, Y.: Fake news detection based on text-modal dominance and fusing multiple multi-model clues. Comput. Mater. Continua **78**(3), 4399–4416 (2024). https://doi.org/10.32604/cmc.2024.047053

17. Mohan, J., Sachin Kumar, S., Soman, K.P.: Synergistic detection of multimodal fake news leveraging TextGCN and vision transformer. Procedia Comput. Sci. **235**, 142–151 (2024). https://doi.org/10.1016/j.procs.2024.04.017

18. Yan, F., Zhang, M., Wei, B., Ren, K., Jiang, W.: SARD: fake news detection based on CLIP contrastive learning and multimodal semantic alignment. J. King Saud Univ. – Comput. Inform. Sci. **36**(8), 102160 (2024). https://doi.org/10.1016/j.jksuci.2024.102160

19. Shan, F., Sun, H., Wang, M.: Multimodal social media fake news detection based on similarity inference and adversarial networks. Comput. Mater. Continua **79**(1), 581–605 (2024). https://doi.org/10.32604/cmc.2024.046202

Open Access This chapter is licensed under the terms of the Creative Commons Attribution-NonCommercial-NoDerivatives 4.0 International License (http://creativecommons.org/licenses/by-nc-nd/4.0/), which permits any noncommercial use, sharing, distribution and reproduction in any medium or format, as long as you give appropriate credit to the original author(s) and the source, provide a link to the Creative Commons license and indicate if you modified the licensed material. You do not have permission under this license to share adapted material derived from this chapter or parts of it.

The images or other third party material in this chapter are included in the chapter's Creative Commons license, unless indicated otherwise in a credit line to the material. If material is not included in the chapter's Creative Commons license and your intended use is not permitted by statutory regulation or exceeds the permitted use, you will need to obtain permission directly from the copyright holder.

Comparative Analysis of Deep Learning Models Using Remote Sensing Image Classification

P. Anushia[1], B. Surendiran[2], B. Prema Mayudu[1],
P. V. S. S. R. Chandra Mouli[1(✉)], and P. Dhivya Bharathi[2]

[1] Central University of Tamil Nadu, Thiruvarur, Tamil Nadu, India
{premamayudu,chandramouli}@cutn.ac.in
[2] National Institute of Technology Puducherry, Karaikal, Puducherry, India
surendiran@nitpy.ac.in

Abstract. Land Use and Land Cover (LULC) classification of remote sensing images is a challenging task. Land Use represents the land utilized by humans, e.g., agricultural land, urban areas, parks, etc. Land cover represents the physical material present on Earth's surface. This study presents a detailed comparative analysis of three advanced deep learning models, namely fuzzy convolutional neural networks, fuzzy CNN generative adversarial networks integrated with GAN-CNN, and a Hybrid CNN-LSTM Architecture for LULC classification. The models are tested on the UC Merced Land Use Dataset. Each model is examined in terms of its architectural design, training methodology, and evaluation based on metrics such as accuracy, area under the curve (AUC), and confusion matrix. Among these three models experimented, the maximum accuracy is obtained for the GAN-CNN model (93.65%), followed by the Fuzzy CNN with 90.79% and the CNN-LSTM model with 83.81%. The findings highlight the potential of the three mentioned models and highlight the mechanism to overcome the challenges of applying different deep learning approaches to remote sensing tasks, offering insights into their respective strengths and suitability for LULC classification.

Keywords: Remote Sensing · Deep Learning · Fuzzy CNN · GAN · CNN-LSTM · UC Merced Dataset

1 Introduction

Remote sensing plays a vital role in monitoring changes in Earth's surface to assess the environment, disasters, agriculture, urban planning, water resources, climate and etc., without direct contact [1]. It collects data about land, water, and the atmosphere using satellites, aircraft, and drones. It's indispensable for sustainable development, disaster preparedness, and environmental protection. Further, decision-making across every field of human activity, such as remote sensing, is crucial for obtaining accurate, timely, and long-term data. Remote

© The Author(s) 2026
J. C. Bansal et al. (Eds.): SCIS 2025, LNNS 1929, pp. 142–154, 2026.
https://doi.org/10.1007/978-3-032-22911-3_11

sensing technology primarily generates image-based data of Earth's surface [3]. In many application domains, the image-based data is being analyzed and processed for decision-making either manually or using traditional image processing techniques. This study aims to process land- use/land-cover (LUCL) data using remote sensing and find how humans are utilizing the land using deep learning techniques [8]. In recent days, artificial Intelligence and deep learning are the most prominent methodologies/techniques to process image data. Computer Vision using Deep learning is highly successful in image classification, segmentation, object detection, high- resolution imaging, and related tasks. [4] In particular, the Convolutional Neural Network (CNN) architectures are most successful in image classification. Classic CNN architectures capture neither contextual ambiguity nor spatial ambiguity, and they are unable to capture some critical temporal dimensions, all of which are important for LULC Classification. To address this issue in the CNN architectures, this work investigates the three combinations with CNN to improve the accuracy and generalization capabilities on LUCL remote sensing data for classification. They are (i) CNN with LSTM to capture the spatial features and temporal dynamics, (ii) CNN with GAN to synthesize the original to increase the data to leverage learning and model expansion, (iii) CNN with Fuzzy logic to address the uncertainty and boundary vagueness.

2   Related Work

2.1   Deep Learning in Remote Sensing

Deep learning methods are suitable for automatic feature selection and more precise analysis of remote sensing images. These methods are more sophisticated when compared with traditional methods. More specifically, CNNs have shown many disadvantages in complex data analyses and processing. Remote sensing images present challenges such as class imbalance (certain land cover types, such as urban areas, are overrepresented), inter-class similarity (sometimes barren lands and dry fields are visually similar), and noise (atmospheric, cloud, and sensor noise can also introduce interference). These issues are addressed through CNN architectures combined with LSTMs (integrating spatial and temporal dimensions), GANs (synthesizing samples to address data imbalance), and Fuzzy logic (managing uncertainty in mixed pixels).

2.2   CNN-LSTM Hybrids

This architecture is used to process both spatial and temporal information on remote sensing data. CNNs process input images to extract spatial features, such as edges, textures, and patterns, present in land cover images [10]. LSTMs analyze sequences and temporal aspects of time- series remote sensing data, such as vegetation growth, climate variations, and land use changes. The combined architecture effectively extracts contextual and dynamic relationships. Hence, it is an optimal choice for this application. [4].

2.3   Generative Adversarial Networks

In a Generative Adversarial Network (GAN), the generator and discriminator are in a competitive game. Each model tries to outdo the other. In remote sensing, GANs mitigate the problems of class imbalance and insufficient labeled data by producing realistic synthetic samples that augment the training data set. The combination of GANs and CNN classifiers improves the management of underrepresented land cover types and rare classes, subsequently enhancing the CNN model's generalization and robustness. CNNs are robust for remote sensing applications, especially for land cover types and healthy under-sampled classes, which are also powerful remote sensing classifiers. [7].

2.4   Fuzzy CNNs

By incorporating fuzzy CNNs into remote sensing applications, convolutional learning can be configured to model and learn uncertainty, noise, and ambiguity. This is useful when land cover areas need to be automatically segmented, especially when boundaries are fuzzy. In uncertain cases, and when fuzzy or mixed pixels are present, soft decision-making can be beneficial. Fuzzy CNNs are especially useful as remote sensing classifiers.

3   Methdology

3.1   Dataset and Preprocessing

The widely used benchmark dataset for remote sensing images is the UC Merced Land Use Dataset. Because of its balanced class distribution, high-resolution imagery, and diverse land- use types. This dataset contains 2,100 images [12]. Each image is 256x256 pixels. The dataset contains 21 different land use categories. Each category consists of 100 images. The data set has diverse landscapes and textures like residential zones, industrial regions, agricultural fields,transport facilities, and others.

3.2   Preprocessing

Prior to model training, all images underwent standard preprocessing procedures. For the CNN-LSTM and GAN-CNN models, images were resized to 128×128 pixels, while for the Fuzzy CNN, a size of 224×224 pixels was used to match the input requirements. Pixel values were normalized to the range [0, 1] to improve training stability. Ground truth labels were one-hot encoded to facilitate multi-class classification. 70% of data is used for training, 15% for validation and 15% for testing.

3.3   Data Augmentation

Data augmentation is used to improve model performance and reduce overfitting. The augmentation pipeline included random rotations, zooming, width and height shifts, and horizontal and vertical flipping. These transformations simulated real-world variations and improved the robustness of the models against spatial and orientation changes in aerial imagery. Figure 1 shows the workflow diagram.

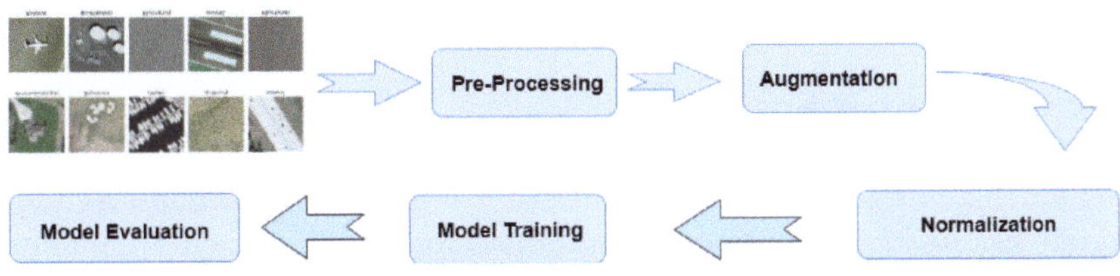

Fig. 1. Workflow Diagram.

4   Deep Learning Architectures

4.1   CNN + LSTM Model

The CNN-LSTM model utilizes MobileNetV2 as the backbone for spatial feature extraction. A global average pooling layer reduces feature dimensions, followed by reshaping the output into a temporal format. This reshaped sequence is passed to a Bidirectional LSTM to capture forward and backward dependencies. The output is then flattened and connected to fully connected layers with ReLU activation, followed by a softmax layer for final classification. This architecture is effective for learning spatial-temporal patterns in aerial image.

4.2   GAN + CNN Framework

The GAN-CNN model integrates a Generative Adversarial Network with a CNN-based classifier. The generator comprises a multi-layer perceptron followed by upsampling layers to produce synthetic images, while the discriminator distinguishes between real and fake images. The synthetic data is combined with the original dataset to improve training diversity. The classification network uses a pre-trained ResNet50 with frozen convolutional layers and custom dense layers for multi-class prediction. This approach addresses data scarcity and class imbalance effectively.

4.3 Fuzzy CNN

The Fuzzy CNN model incorporates fuzzy logic into the CNN architecture to model uncertainty in remote sensing imagery. MobileNetV2 serves as the backbone for initial feature extraction. A fuzzy layer is introduced, where the ReLU-activated feature map is element-wise multiplied with a sigmoid-based fuzzy activation map, enhancing feature discriminability. This is followed by pooling and dropout layers to reduce overfitting, and final dense layers lead to a softmax output. This structure improves performance in regions with ambiguous boundaries.

5 Experimental Setup

TensorFlow 2.x is used for implementation. The learning rate is set as 1×10^{-4} and batch size as 32. Each model was trained for 5 to 15 epochs, depending on convergence behavior. Early stopping was employed based on validation loss to avoid overfitting. Data augmentation and preprocessing were performed on-the-fly during training. Evaluation was conducted using accuracy, AUC, confusion matrix, and classification report [11].

6 Results and Performance Analysis

6.1 Quantitative Evaluation

Table 1. Accuracy Comparison of Models

Model	Train Accuracy (%)	Validation Accuracy (%)	Test Accuracy (%)
CNN + LSTM	80.88%	81.59%	83.81%
GAN + CNN	99.66%	95.24%	93.65%
Fuzzy CNN	97.62%	92.38%	90.79%

Table 1 presents the training, validation, and testing accuracies of the three models. The testing results of three models are summarized in Table 1. The CNN combined with the GAN model noted the highest training accuracy (99.66 test accuracy compared with others. The other two models, such as CNN combined with LSTM and Fuzzy logic, showed adequate results. Table 1 presents the comparative results of three models.

6.2 Visualizations and Metrics

The training accuracy and loss for all models over the epochs are illustrated in
Figs. 2, 3, and 4. We also offer a comparison of image classification results for
the 21 land use classes across the three models. Moreover, the training accuracy
and loss curves show the convergence and stability of the models which are given
in Figs. 5 and 6 respectively.

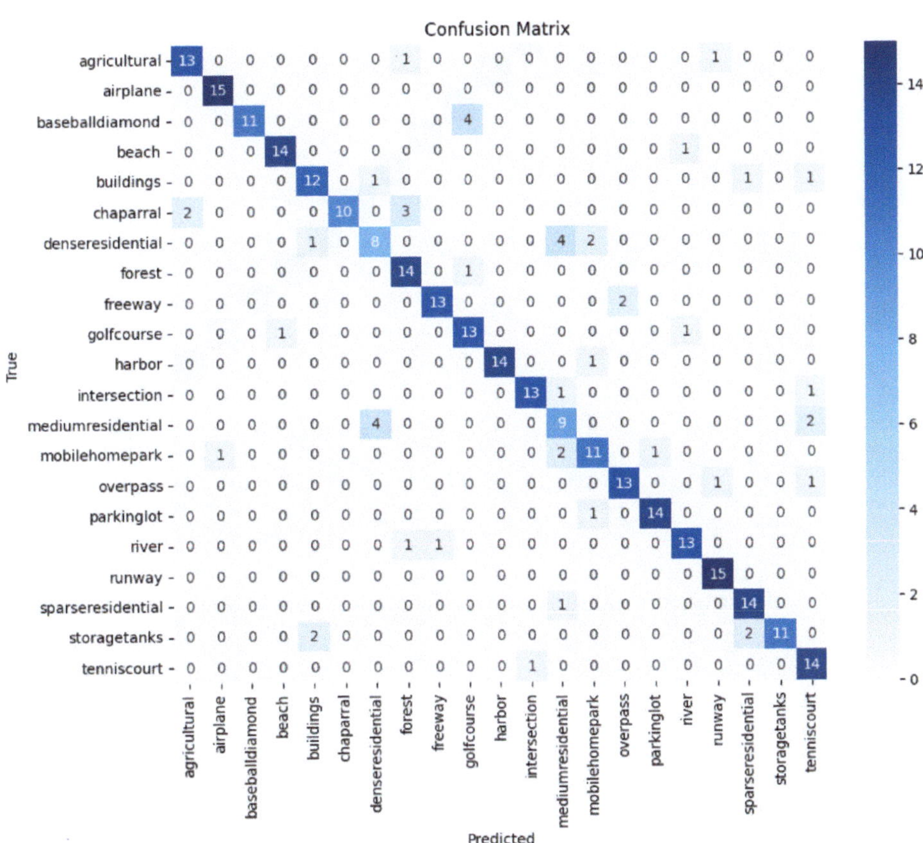

Fig. 2. Confusion Matrix for CNN+LSTM Model

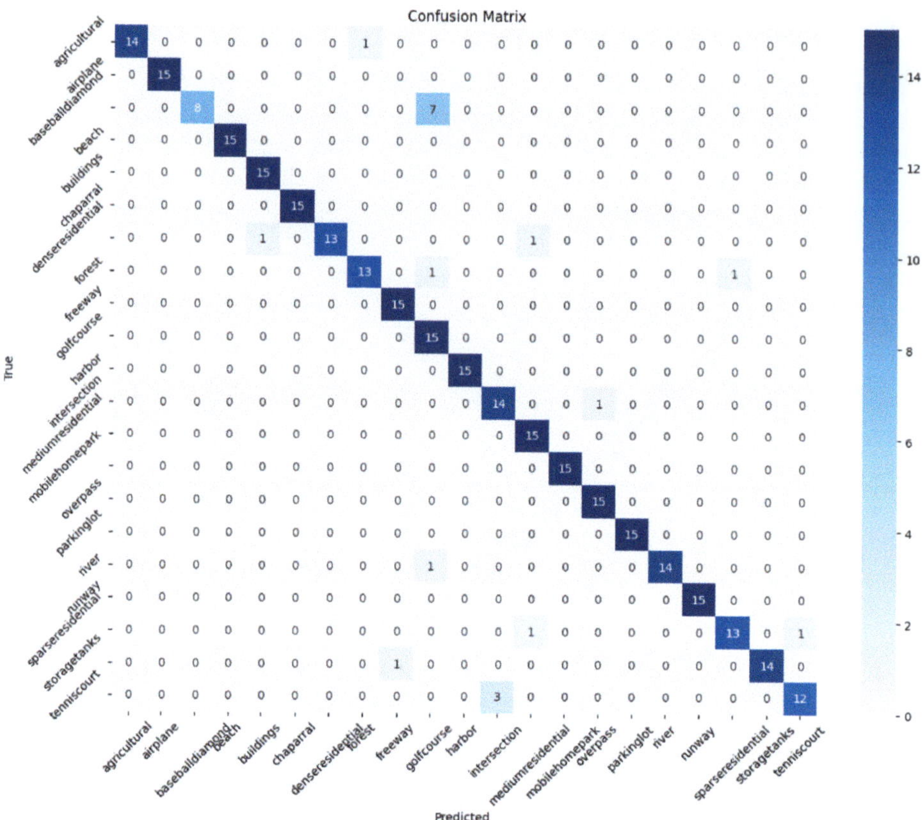

Fig. 3. Confusion Matrix for GAN+CNN Model

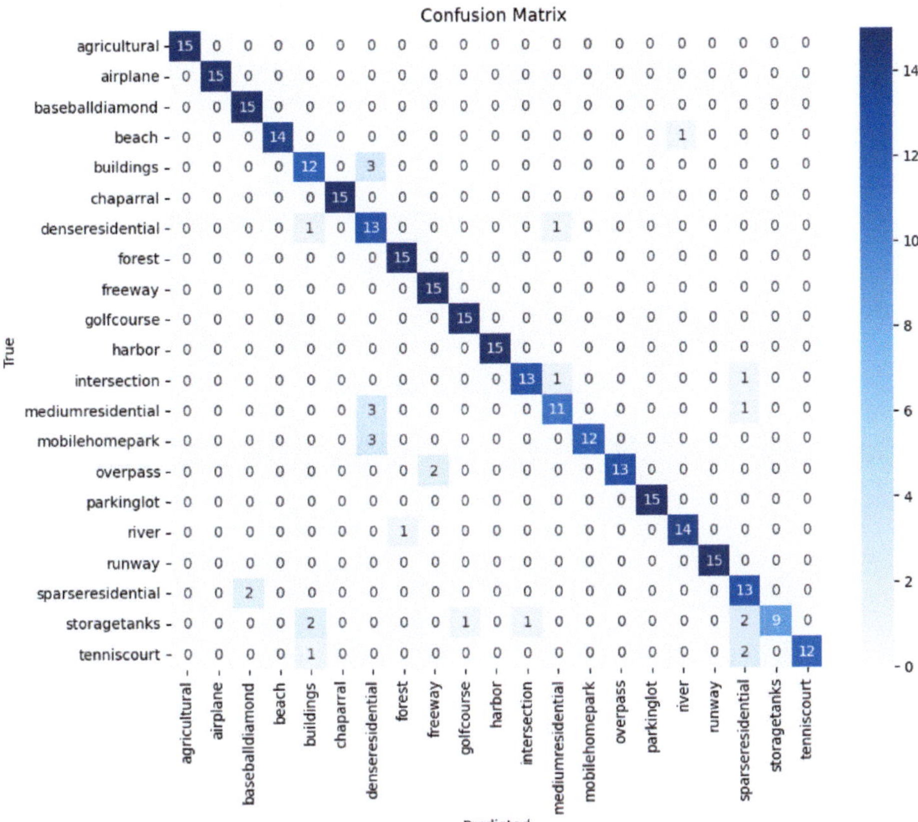

Fig. 4. Confusion Matrix for Fuzzy CNN Model

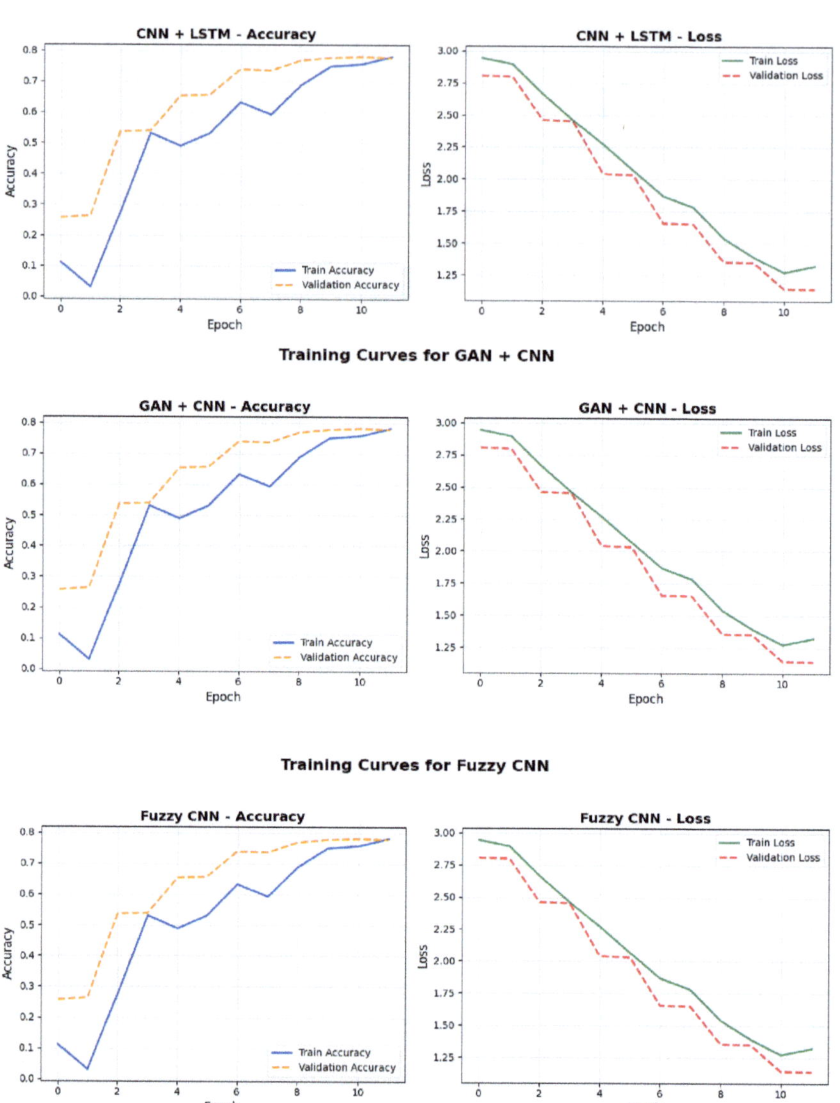

Fig. 5. Training Accuracy and Loss Over Epochs for CNN-LSTM, GAN-CNN, and Fuzzy CNN

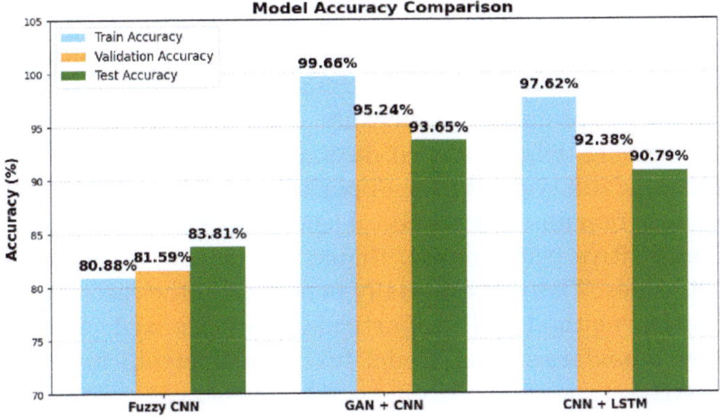

Fig. 6. Comparison of accuracy over the three models

7 Model Strengths

Among all models, the GAN-CNN is the one that performed the best, generating synthetic samples that expanded the training dataset and helped to alleviate class imbalance. The Fuzzy CNN reached high classification accuracy because it captured the uncertainty and boundary ambiguity that characterizes remote sensing images. The CNN-LSTM model captured the sequential and contextual dependencies, yet its temporal modeling was of little use with a purely spatial dataset.

7.1 Challenges

The GAN-CNN framework, although performing well, is still exposed to mode collapse, which results in a lack of diversity within the synthetic samples. The performance of the CNN-LSTM [5] model is highly dependent on the chosen sequence length, which is very likely to lead to overfitting if left unregulated. The Fuzzy CNN has to be very specific in the calibration of its fuzzy activation, especially concerning the interaction of the sigmoid and ReLU functions, which could then lead to divergence.

8 Discussion

The challenges and limitations of three models, such as CNN+LSTM, CNN+GAN, and CNN+Fuzzy, on the UC Merced Land Use Dataset are provided in the comparative evaluation. The CNN+LSTM model demonstrates its ability to capture spatio-temporal dependencies. This enabled the model to capture contextual variations across diverse land cover types. However, this model requires significant computational resources and is computationally intensive. In addition, it declines with high intra-class similarity. In contrast, the

CNN+GAN model addressed class imbalance and reduced overfitting. This model achieved the highest accuracy and Area Under the Curve (AUC) among the three. It also enhanced the model's robustness and generalization.

The results were promising, especially given the use of adversarial learning on remote sensing, where labeled data and/or class distribution are imbalanced. Training instability due to the opposing dynamics of the GAN's generator and discriminator remains a downside. Lastly, the Fuzzy CNN model similarly achieved a competitive performance, thanks to the convolutional integration of fuzzy logic principles. This was especially beneficial in the cases where traditional crisp classification methods failed, particularly with mixed pixels and ambiguous land cover boundaries. The model fuzzy membership functions addressed the uncertainty.

This approach has merits; however, the fuzzy-based model seems most useful when uncertainty is very high, since its accuracy is lower than that of the GAN-CNN. Hence, fuzzy-based models may only be useful in high-uncertainty scenarios as opposed to all-class situations. This indicates that no one architecture is the best for all situations. Rather, it should be aligned to the specific type of classification task. For instance, the CNN-LSTM is most suitable for tasks that need sequence and contextual information, situations that are GAN-CNN favorable are those that are data-scarce and imbalanced, while the Fuzzy CNN is suitable for situations that have uncertainty. This also suggests the need for hybrid frameworks incorporating temporal modeling, adversarial learning, and fuzzy logic to deal with more sophisticated problems in land-use and land-cover classification. Adopting a largely temporal pattern or model will most likely constrain the potential performance.

9   Conclusion and Future Work

This study evaluated and contrasted the performance of three deep learning architectures—CNN- LSTM, GAN-CNN, and Fuzzy CNN—specifically for land-use and land-cover classification with the UC Merced Land Use Dataset. Overall, the GAN-CNN model classifier demonstrates the most significance of synthetic data augmentation, and the highest classification accuracy as well as the highest AUC. The Fuzzy CNN model, however, had the highest performance with indistinct boundaries. The CNN-LSTM model [9], on the other hand, was the most underwhelming in a largely spatial setting, which was the most pivotal for capturing sequential elements of a given context [2]. Developing hybrid frameworks for the seamless integration of the boundary- enhancing features of Fuzzy Logic and the Generative learning encompassed within Fuzzy- GAN-type models, along with the data diversity constructs offered by Fuzzy-CNN, will be the primary focus of research. Moreover, the combination of LSTM and Fuzzy system will create new deep learning opportunities in Remote Sensing for spatial uncertainty problems, leading Remote Sensing to the next phase [6].

References

1. Cheng, B., Li, Z., Xu, B., Dang, C., Deng, J.: Target detection in remote sensing image based on object-and-scene context constrained CNN. IEEE Geosci. Remote Sens. Lett. **19**, 1–5 (2022). https://doi.org/10.1109/LGRS.2021.3087597
2. Gupta, S., Dwivedi, R.K., Kumar, V., Jain, R., Jain, S., Singh, M.: Remote sensing image classification using deep learning. In: 2021 10th International Conference on System Modeling Advancement in Research Trends (SMART), pp. 274–279 (2021). https://doi.org/10.1109/SMART52563.2021.9676295
3. Hussein, L., Prabhavathi, V., Patil, R., Nagendar, Y., Devi, T.A.: Satellite image-based flood monitoring and prediction using multi verse optimization algorithm with support vector machine. In: 2024 International Conference on Distributed Systems, Computer Networks and Cybersecurity (ICDSCNC), pp. 1–5 (2024). 10.1109/ICDSCNC62492.2024.10939213
4. Narayana, V.L., Bhargavi, S., Srilakshmi, D., Annapurna, V., Akhila, D.M.: Enhancing remote sensing object detection with a hybrid densenet-LSTM model. In: 2024 IEEE International Conference on Computing, Power and Communication Technologies (IC2PCT). vol. 5, pp. 264–269 (2024). https://doi.org/10.1109/IC2PCT60090.2024.10486394
5. Patel, S., Ganatra, N., Patel, R.: Multi-level feature extraction for automated land cover classification using deep CNN with long short-term memory network. In: 2022 6th International Conference on Trends in Electronics and Informatics (ICOEI), pp. 1123–1128 (2022). https://doi.org/10.1109/ICOEI53556.2022.9777148
6. Saranya, K., Bhuvaneswari K, S.: Semantic annotation of land cover remote sensing images using fuzzy CNN. Intell. Automation Soft Comput. **33**, 399–414 (2022). https://doi.org/10.32604/iasc.2022.0231497
7. Shi, C., Fang, L., Lv, Z., Shen, H.: Improved generative adversarial networks for VHR remote sensing image classification. IEEE Geosci. Remote Sens. Lett. **19**, 1–5 (2022). https://doi.org/10.1109/LGRS.2020.3025099
8. T V, V., C, I.: Remote sensing image captioning using CNN and LSTM. In: 2024 International Conference on Advancements in Power, Communication and Intelligent Systems (APCI), pp. 1–6 (2024). https://doi.org/10.1109/APCI61480.2024.10616900
9. Tulapurkar, H., Banerjee, B., Mohan, B.K.: Effective and efficient dimensionality reduction of hyperspectral image using CNN and LSTM network. In: 2020 IEEE India Geoscience and Remote Sensing Symposium (InGARSS), pp. 213–216 (2020). https://doi.org/10.1109/InGARSS48198.2020.9358957
10. Varma, B., Naik, N., Chandrasekaran, K., Venkatesan, M., Rajan, J.: Forecasting land-use and land-cover change using hybrid CNN–LSTM model. IEEE Geosci. Remote Sens. Lett. **21**, 1–5 (2024). https://doi.org/10.1109/LGRS.2024.3389671
11. Wang, J., et al.: Deep hierarchical representation and segmentation of high resolution remote sensing images. In: 2015 IEEE International Geoscience and Remote Sensing Symposium (IGARSS), pp. 4320–4323 (2015). https://doi.org/10.1109/IGARSS.2015.7326782
12. Yang, Y., Newsam, S.: Bag-of-visual-words and spatial extensions for land-use classification. In: Proceedings of the 18th SIGSPATIAL International Conference on Advances in Geographic Information Systems, pp. 270–279 (2010)

Open Access This chapter is licensed under the terms of the Creative Commons Attribution-NonCommercial-NoDerivatives 4.0 International License (http://creativecommons.org/licenses/by-nc-nd/4.0/), which permits any noncommercial use, sharing, distribution and reproduction in any medium or format, as long as you give appropriate credit to the original author(s) and the source, provide a link to the Creative Commons license and indicate if you modified the licensed material. You do not have permission under this license to share adapted material derived from this chapter or parts of it.

The images or other third party material in this chapter are included in the chapter's Creative Commons license, unless indicated otherwise in a credit line to the material. If material is not included in the chapter's Creative Commons license and your intended use is not permitted by statutory regulation or exceeds the permitted use, you will need to obtain permission directly from the copyright holder.

Adaptive Augmentation and Synthetic Image Generation (AAg-SiG) Framework for Food Applications

Dilpreet Singh Brar[1] [iD], Indu Bala[2(✉)] [iD], Birmohan Singh[3] [iD], and Vikas Nanda[1] [iD]

[1] Department of Food Engineering and Technology, Sant Longowal Institute of Engineering and Technology, Longowal, Punjab 148106, India
`vikasnanda@sliet.ac.in`
[2] School of Computer and Mathematical Sciences, Faculty of Sciences, Engineering and Technology, University of Adelaide, Adelaide, Australia
`iinduyadav@gmail.com`
[3] Department of Computer Science and Engineering, Sant Longowal Institute of Engineering and Technology, Longowal, Punjab 148106, India

Abstract. This work developed a novel dataset generation framework based on Adaptive Augmentation and Synthetic Image Generation (AAg-SiG), which improves the image quality in the original dataset and generates synthetic images to balance the input dataset. The proposed framework works on the class-based feature adaptation method to fix anomalies in the images, like brightness, colour variations, blur using Gaussian Blur, CLAHE and UnsharpMark methods. The Oriented FAST and Rotated BRIEF (ORB) technique was used for geometric augmentation by using controlled affine transformation and key point detection. Moreover, the class mean (soft colour alignment) was used to fix the RGB distribution of the image, which preserved the specific features of all the images in the respective dataset class. Additionally, the interquartile range (IQR) was used to generate the class-specific synthetic image by calculating RGB channel features. Besides, the final stage was the validation of the generated image; an input class-based pre-fixed quality metrics were used to compare the augmented and generated image for authentication. Furthermore, the evaluated ORB heatmaps results justify the effectiveness of the AAG-SiG framework, as the important features (i.e. edges and feature density) were presented in the generated images. All in all, the AAg-SiG can be a useful dataset augmentation and generation tool to integrate with real-world applications of AI-models in the food process industry for strengthening the intelligent quality assurance (QA) and quality control (QC) systems.

Keywords: Image generation · Adaptive augmentation · Food Quality evaluation · Rotated BRIEF (ORB) · Oriented FAST and Red chilli powder

© The Author(s) 2026
J. C. Bansal et al. (Eds.): SCIS 2025, LNNS 1929, pp. 155–167, 2026.
https://doi.org/10.1007/978-3-032-22911-3_12

1 Introduction

The application of AI-technology in the food quality assessment is solely dependent on the training dataset quality, variability, and authenticity [1]. The digital image-based methods provide a scaling and non-destructive alternative to the conventional technology to detect the adulteration in the red chilli powder (RCP) [2]. Although databases are often acquired through controlled variables (settings), they often have limitations in the form of blurriness, unstable brightness, reflections, and uneven luminance. These might not only undermine the visual quality, but also the ability of the AI-model to generalise or classify accurately [3].

Various image augmentation methods are developed, which include fundamental image transformations like photometric transformation (i.e. contrast, hue, saturation and brightness) and geometric processes (i.e. flipping, rotation, scaling and cropping) [4, 5]. Nevertheless, such approaches do not always cope with the perceptual quality, which is important in extracting and classifying features [4]. The above-discussed methods are applied to improve the dataset quality in terms of heterogeneity to improve the robustness and generalisation ability by reducing the overfitting in the AI-model. Along with more traditional augmentation methods, more advanced approaches have been developed, including diffusion-based generative models, Generative Adversarial Networks (GANs), and Variational Autoencoders (VAEs) that have become effective technologies in the generation of synthetic datasets [6]. These have been used in food imaging to add more data and also enhance the quality estimation, anomaly detection and classification problems [7]. In spite of such progress, especially when it comes to food image applications, GANs are also faced with the hindrances of training instability, high computing cost and model collapse [8–10]. In addition, such models tend to fail to capture the fine-scale texture, reducing their usefulness in creating a realistic food dataset [11]. Hence, as a way of mitigating the drawbacks of the existing image augmentation and synthetic image methods in the context of RCP, an adaptive framework for image augmentation is suggested. Such an approach retains the most relevant information in the product in the sample image, and hence it helps the AI-model to acquire important features and accurately predict the adulteration in the food product.

To cover the existing gap related to adaptive augmentation and generation methods with specific food applications. This research aims to develop an adaptive augmentation and synthetic image generation (AAg-SiG) method, which addresses the challenges in the application of the image RcP dataset. The proposed novel framework assesses image quality, applies target enhancements, and ensures class-specific colour consistency. Furthermore, it introduces orientation-aware geometric variations using oriented FAST and Rotated BRIEF (ORB) and generates synthetic images within class-specific RGB shift limits. This framework offers significant applications in food quality evaluation and adulteration detection in powders and granular food products such as turmeric, RcP and other spice blends. By improving dataset diversity while maintaining perceptual and class-specific integrity, this framework enhances the training of robust deep learning models for destructive food quality analyses, time food sorting systems and supply chain monitoring and real.

2   Materials and Methods

2.1. Framework: The proposed work developed a framework (AAg-SiG) for correcting RcP images using adaptive augmentation, utilising the interquartile range and generating synthetic images with the ORB technique (Fig. 1).

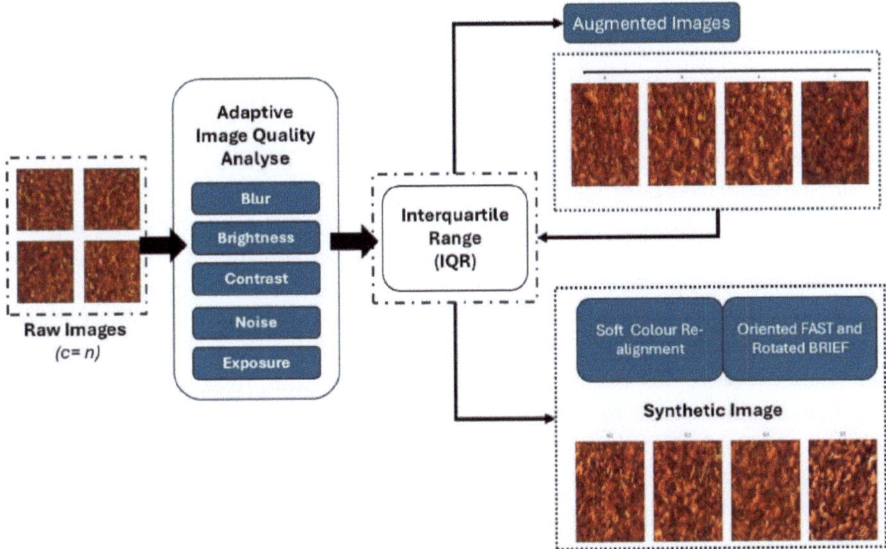

Fig. 1. Farmwork of adaptive augmentation and synthetic image generation (AAg-SiG)

2.2. Dataset: The digital image dataset was developed by following the methodology proposed by Brar et al. [11]. It consists of a total of 20 classes, from which six classes correspond to Pure RCP samples (three high-grade and three low-grade). Besides, 15 classes represented the adulteration of low-grade RCP (varieties - BM, UT and GM) with Sudan(IV) at varying concentrations (0.1%, 0.2%, 0.4%, 0.6% and 1.0%).

2.1   Augmentation and Generation of Synthetic Images

Image augmentation for correction of imperfections involves applying targeted transformations, such as deblurring, denoising, contrast adjustment, and exposure correction, to enhance image quality. The AAg-SiG approach aims to rectify errors introduced by human or instrumental factors during image acquisition, thereby ensuring cleaner and more consistent data for analysis or model training. The synthetic image generation process through the Oriented FAST and Rotated BRIEF (ORB) algorithm follows the model of salient keypoint identification and retrieving the orientation-sensitive description of raw image data. The learnt description then guides the parametric manipulations such as scaling, wrapping and rotating, which are then used to generate new synthetic samples that maintain the structural identity of the objects of interest but increase sample variety in the database [12]. Subsequently, this method enhances the stability of the models with respect to spatial distortions. Initially, each image was evaluated for quality using perceptual metrics calculated on its greyscale equivalent, which was obtained from RGB using standard luminance conversion. These measurements include brightness, colour cast, blurriness, noise, and specular reflection (proportion of pure white pixels). Defective photographs were subjected to an adaptive correction module that includes Gaussian Blur, Unsharp Masking, and CLAHE filters based on this assessment. A soft colour-matching stage then uses a weighted scaling method to gently push the corrected image's RGB distribution in the direction of the mean RGB vector of its related class.

In order to generate synthetic images, a randomised angle perturbation and affine rotation are performed after introducing geometric variation, utilising ORB key points to assess dominant orientation. The IQR of each class's colour distribution was adaptively used to determine the limits of brightness-contrast adjustments and class-specific RGB shifts that were applied to this rotated image. Every synthetic image was assessed against predetermined metric bounds, which were determined using the original dataset's IQR and tolerance thresholds, in order to guarantee realism. Only when all metrics fall within these bounds was the image considered acceptable. In addition to improving the resilience of downstream models, this quality gate guaranteed that augmented and synthetic images maintain their statistical alignment with real samples by enforcing consistency in blurriness, exposure, colour fidelity, and other perceptual aspects (Fig. 1). The mathematical representation of the framework (AAg-SiG) is as follows.

2.1.1 Image Quality Metrics

The quality metrics of each image in the dataset were evaluated. To analyse the image properties, the RGB image (I_R, I_G, I_B) was converted into grayscale following Eq. (1) [13].

$$I_g = 0.299.I_R + 0.587.I_G + 0.114.I_B$$

Here, I_g = image, using Eq. (1), the following perceptual metrics are defined:

a). **Blurriness:** Variance of the Laplacian of the grayscale image (Eq. 2):

$$M_{blur}(I) = Var(\nabla^2 I_g) \tag{2}$$

where $M_{blur}(I)$ represents the blurriness of image I_g.

b) **Noise:** The noise for the image I_g can be defined as:

$$M_{noise}(I) = Std(\nabla^2 I_g) \tag{3}$$

where $Std(\nabla^2 I_g)$ is the standard deviation for the Laplacian of I_g.

c) **Brightness**: it is calculated as the Mean intensity of I_g as follows (Eq. 4):

$$M_{brightness}(I) = \mu(I_g) \tag{4}$$

d) **Exposure:** Pixels under- or over-exposed (Eq. 5):

$$M_{\text{exposure}}(I) = \frac{C\left(I_g < 15 \vee I_g > 240\right)}{H \cdot W} \tag{5}$$

where, C denotes the cardinality of pixels, W, H, W are width and the height of the pixels in I_g, respectively, $\vee$, represents *'logical-OR'* operator.

e) **Colour Cast:** Relative deviation among mean RGB channels (Eq. 6):

$$M_{\text{cast}}(I) = \frac{\text{Std}(\mu_{RGB})}{\mu(\mu_{RGB}) + \epsilon} \tag{6}$$

where $\epsilon > 0$ and $\epsilon \to 0$, $\mu_{RGB} = \{\mu_R, \mu_G, \mu_B\}$

f) **Specular Reflection:** Fraction of pure white pixels (Eq. 7):

$$M_{\text{specular}}(I) = \frac{C(IG > 240)}{H \cdot W} \tag{7}$$

where, $IG = \{i_g; i_g \epsilon \Psi\}$, s.t. $\Psi = \{I_R, I_G, I_B\}$

2.1.2　Adaptive RGB Shift Limits

For each class c, the mean values for each channel (i.e. R, G, B) across all its images can be represented as $\mu_c(R)$, $\mu_c(G)$, $\mu_c(B)$. Estimating the interquartile ranges (IQRs) [14] for each class in each channel as follows

$$\text{IQR}_c(x) = Q_3^x - Q_1^x; x \in \{R, G, B\} \tag{8}$$

Further, an adaptive RGB shift limit for each image was assigned:

$$\Delta_c(x) = \text{clip}(0.75 . \text{IQR}_c(x); \textit{min} = 2, \textit{max} = 15) \tag{9}$$

These were used in synthetic image augmentation to match the class-specific colour variability.

Adaptive Correction with Soft Colour Matching Images identified as defective were enhanced using a static filter pipeline $\mathcal{T}_{\text{augM}}$ consisting of CLAHE, UnshrpMask [15], and GaussianBlur [16].

Let: $I' = \mathcal{T}_{\text{augM}}(I_g)$, where $\mathcal{T}_{\text{augM}}(I_g)$, is a static filter pipeline of an image I_g, consisting of three filters (mentioned above).

To softly align the image with the class mean $\mu_c \in \mathbb{R}^3$, ($\mu_c = (\mu_c(R), \mu_c(G), \mu_c(B))$ is derived from Eq. 8).

Applying channel-wise scaling.

$$I_{\text{fixed}} = \text{clip}\left(I' \cdot \left(\alpha \cdot \frac{\mu_t}{\mu_{I'}} + (1 - \alpha)\right), 0,255\right) \tag{10}$$

where μ_t is the target mean of the RGB vector $((\mu_c(R), \mu_c(G), \mu_c(B))$, $\mu_{I'}$ is the mean RGB of the transformed image and α is a tuning parameter which is set to 0.85.

Synthetic Image Generation Using ORB-Based Rotation To introduce geometric variation, ORB key points are used to estimate dominant orientation [17]:

$$\theta = \text{median}\left(\{\theta_k\}_{k=1}^{K}\right) + \mathcal{U}\left(-25°, +25°\right) \tag{11}$$

where, θ is the rotation angle on images I_g, I, and I_{fixed} determined from ORB key points. The image is then rotated using I_{rot} s.t.

$$I_{rot} = \text{warpAffined}\ (I, \theta) \tag{12}$$

where, warpAffined (I, θ) is the image transformation function of the image I with respect to θ. It is followed by RGB augmentation with limits derived from Sect. 2 by following the equation:

$$I_{Syn} = \mathcal{T}_{Syn}(I_{rot}) \tag{13}$$

where, $\mathcal{T}_{Syn}$ includes RandomBrightnessContrast and class-specific RGBShift of the synthetic image I_{Syn}.

Quality Gate for Synthetic Validation To establish the effectiveness of the generated synthetic images are statistically similar to the original dataset, the per-metric bounds are defined as follows:

For each metric m:

$$\text{Bounds}_m = \left[(Q_1^m - \tau \cdot \text{IQR}_m), (Q_3^m + \tau \cdot \text{IQR}_m)\right] \tag{14}$$

where, $\tau = 0.25$ is the tolerance.

A synthetic image I_{Syn} was accepted only if:

$$\forall m, M_m\left(I_{syn}\right) \in \text{Bounds}_m \tag{15}$$

where, $M_m\left(I_{syn}\right)$, is the computed score of the metric m, on the synthetic image I_{syn}.

2.2 Computational Specifications:

The experiments are performed using Python 3.10, Pytorch 2.1 and Albumenations on an Intel Xeon-4215 processor and a NVIDIA A6000 GPU (48 GB).

3 Results and Discussion

The application of AI technology has been rapidly expanding across various domains of the food industry, encompassing areas such as food quality evaluation, supply chain management, process optimisation, tracing foreign substances, precision fermentation, etc. The reliability and effectiveness of AI-based solutions are intrinsically linked to the quality and structure of the dataset on which they are trained and deployed [1]. The data and the AI-model in this case are symbiotic in nature; diversity and quality are the basis of creating strong and generalizable AI-systems with dataset-dependent decision making and functionality. When it comes to the classification of food products, especially powdered products in the image, many multidimensional factors may come into play to determine the final results of the classifier. Environmental conditions during image acquisition, such as illumination, temperature and relative humidity (RH) is one of the major sources of variability [12]. In order to reduce this variability, the best solution will be to take pictures in a controlled environmental background with stable lighting, temperature, and RH. Nevertheless, human and instrumental error effects may be associated with the artefacts of blur, over- or under-exposure, shadows, and uneven brightness, even in controlled settings, all of which lead to the adverse effects of classification accuracy of the model. The AAgSiG framework that has been developed is very efficient in order to overcome these challenges.

Based on Fig. 2i, which shows the original images of RcP samples, including those contaminated with natural substances, such representative images have natural colour tones, particle textures, and visual variance that are related to reality. However, there are also pictures with problems like blurriness or out of the normal colour deviations, which may bring noise to the dataset and, thus, can misclassify, which decreases the accuracy of adulteration detection models. In order to solve these exceptions and promote data consistency, an adaptive augmentation strategy, AAgSiG, was enforced. The strategy uses class-specific constraints depending on the IQR of image enhancement parameters that are important: the CLAHE (ContrastLimited Adaptive Histogram Equalisation), UnsharpMask, and GaussianBlur. The image with blurriness or contrast levels outside these stipulated statistical limits is corrected as such. Figure 2ii illustrates the effect of the approach, in particular, the reduction of excessive blurring and normalisation of visual characteristics in all samples. This preprocessing makes the dataset within limited visual limits, hence enhancing the quality and reliability of the input data to be used in downstream classification.

The comparison of the augmented and the Raw RcP images and the respective ORB key-point heatmap illustrated the structural preservation in the augmented images using the AAg-SiG (Fig. 3). The ORB heatmaps of the original pictures have well-spread key points that reflect the natural texture and granularity of the chilli particles. Interestingly, the augmented images maintain and even improve this density of key points, proving that the transformations applied to them (i.e. CLAHE and UnsharpMask) are successful at boosting the local features without corrupting the underlying structure. The improvement and procurement of the ORB key points reflect better textural feature characterisation and edge delineating, which is useful in important feature extraction for precise classification between classes. Moreover, the reproducibility of a number of samples is used to verify the stability and controllability of the augmentation process so that every change in the synthesised image is structurally and visually viable. In general, this analysis using ORBs demonstrates that the AAgSiG strategy is effective in enhancing the dataset and preserving the high level of fidelity to original image features, thus improving the quality and accuracy of AI models for food QA and QC applications.

In addition, Fig. 4 depicts a grid of synthetic RcP images generated by an AAg-SiG structure that combines CLAHE, UnsharpMask, soft colour matching, GaussianBlur, and ORB-guided rotation. Images with visual defects were corrected and made to match

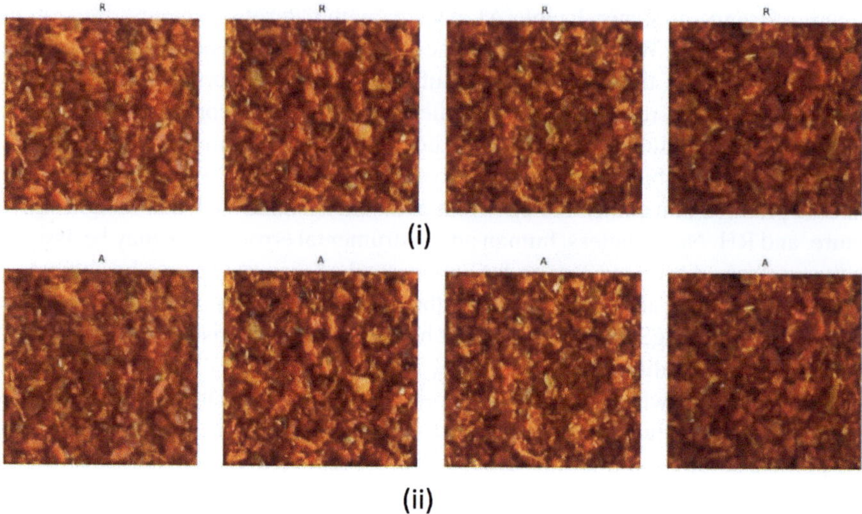

Fig. 2. (i) Raw image of Red chilli Powder (RPC); (ii) Augmented image using AAg-SiG method

the class-specific meanings. The RGB adjustment (RGBShift, RandomBrightnessContrast) and the ORB-controlled geometric diversity ensure stability and variability in the generated image, similar to a raw image. The result is a synthetically rich and aesthetically consistent synthetic dataset that is incredibly useful in terms of improving model robustness of food adulteration classification. As depicted in Fig. 4, images generated maintain the texture of natural grains, even colour distribution and realistic structure. Unlike the traditional methods of augmentation that can add artefacts and weaken the integrity of classes, the suggested framework generates high-quality synthetic images with high levels of intra-class and meaningful inter-sample variation. These properties make the dataset very effective in expanding small datasets or imbalanced datasets, especially when there are fine-grained colour and texture cues that would help achieve accurate results in any classification.

Figure 5 demonstrates a typical synthetic image along with the corresponding ORB key-point heatmap that provides a good understanding of the spatial resolution and the number of features of the created sample. The generated image is highly similar to natural RcP in the aspects of uniformity of texture, transparency of grains and the natural colour shade, which is due to the well-tuned adaptive augmentation pipeline. The associated ORB heatmap displays a high concentration and widespread distribution of key points, reflecting strong structural preservation and retention of fine textural cues vital for classification. The appearance of multiple high-intensity zones in the ORB map confirms the maintenance of edge sharpness and local contrast, which are essential for effective feature extraction in deep learning models. This reinforces the efficacy of the proposed AAg-SiG strategy in not only enhancing visual quality but also in preserving discriminative features necessary for accurate and reliable AI-driven food quality assessment.

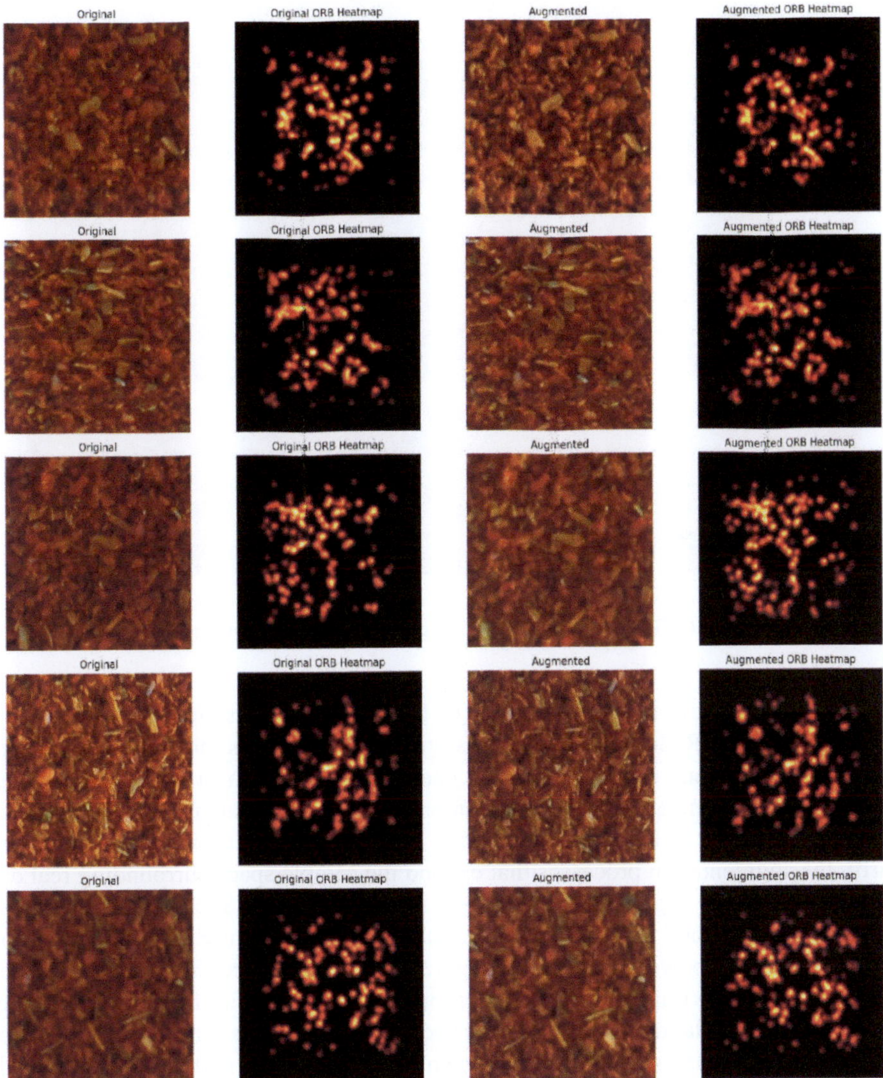

Fig. 3. Heat map of ORB keypoint for the raw image and the Augmented image of Red Chilli powder (RCP)

To date, no standardised method exists for augmenting and generating synthetic images of food powder datasets. This study is the first to introduce an effective adaptive augmentation (AAg-SiG) framework specifically designed for powdered food images. According to existing literature, Xu et al. [4] classified augmentation techniques into three main types: model-based, model-free and policy-optimisation strategies. Model-free methods, like spatial transformations and pixel adjustments, offer simple ways to increase data diversity without requiring models [18]. Model-based approaches, such as GANs, VAEs, and label-conditioned generators, help address data imbalance and domain

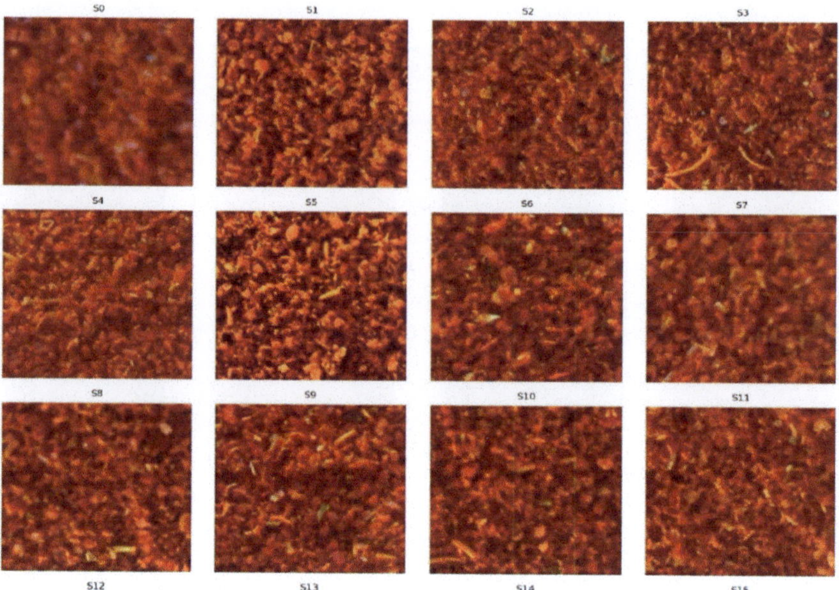

Fig. 4. AAg-SiG Generated Synthetic Images of Red chilli Powder (RCP) samples

shifts by creating new, relevant samples [6, 19]. Policy-optimisation methods, such as AutoAugment and RandAugment, learn augmentation policies through reinforcement or adversarial learning to enhance model generalisation [20]. Yun et al. [21] highlight challenges such as image variation, class imbalance, domain adaptation, and overfitting, which are common in food image datasets. Vicinity distribution is a concept used to clarify the augmentation processes that expand the sample space surrounding real data points and make the augmentation processes more robust [22]. Such methodologies are actively used in natural image processing, but their systematic use on alimentary products, in particular, powdered commodities (red chilli) have been limited. The current study, therefore, fills this gap through incorporating class-specific enhancement and perceptual quality regulation and ORB-based geometric transformations to provide reliable synthetic image samples in developing food quality monitoring systems to deploy in the food industry.

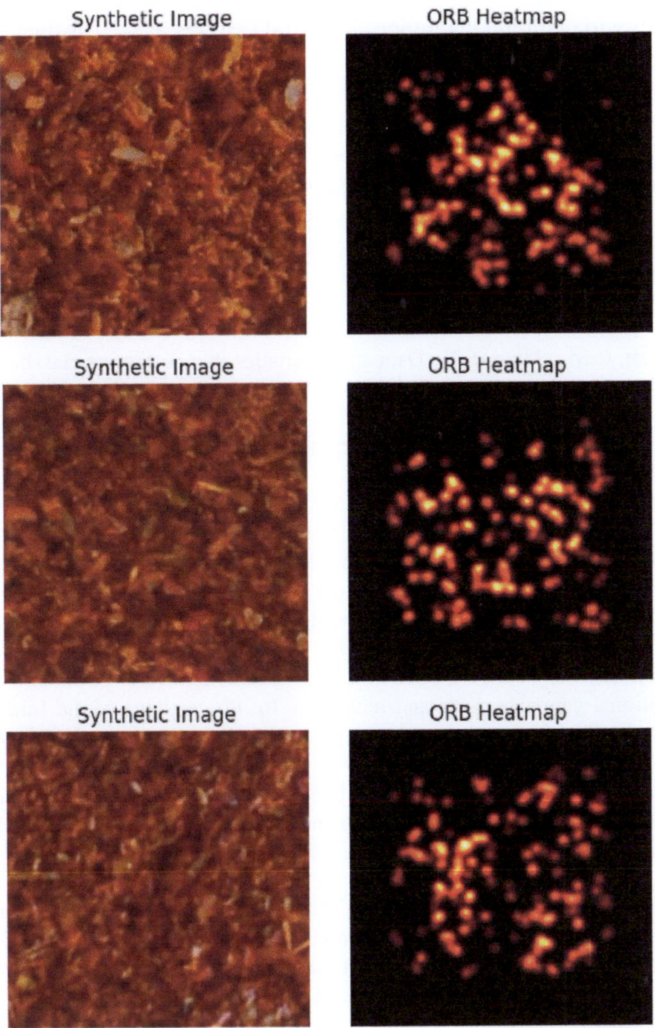

Fig. 5. Heat map of ORB keypoint for generated synthetic images using AAg-SiG

4 Conclusion

The paper presents a highly designed adaptive augmentation (AAgSiG) framework, which is expected to enhance the quality, consistency, and variety of food image data with special emphasis on adulteration in raw consumer products (RcP). The framework combines the methods of enhancing images in a stationary manner, including CLAHE, UnsharpMask, and GaussianBlur, as well as soft colour alignment to the direction of class-specific means and ORB-based rotation to obtain controlled geometric variation.. The components are effective in reducing the typical imaging artefacts like blur, unbalanced lighting, and colour anomalies. This framework scales up the visual quality of the raw and augmented images, and the synthetic images generated through the AAg-SiG have the capacity to maintain the important digital features as well as the inter-class variation. The retention of fine-grained local features that are essential to obtaining accurate classification is supported by ORB keypoint heatmaps. The suggested framework, therefore, enhances the reliability of datasets and the creation of strong and high-performance artificial-intelligence models to estimate food quality. It is also amenable to more generic uses, in the testing of food authenticity, safety and intelligent visual inspection in a wide range of agri-food industries due to its modularity and scalability.

References

1. Van der Walt, C.M., Barnard, E.: Data characteristics that determine classifier performance. SAIEE Africa Res. J. **98**(3), 87–93 (2021)
2. Brar, D.S., Singh, B., Nanda, V.: Application of image-based features and machine learning models to detect brick powder adulteration in red chili powder. J. Food Process Eng **47**(11), e14762 (2024)
3. Chandler, D.M.: Seven challenges in image quality assessment: past, present, and future research. Int. Scholar. Res. Notices **2013**(1), 905685 (2013)
4. Xu, M., Yoon, S., Fuentes, A., Park, D.S.: A comprehensive survey of image augmentation techniques for deep learning. Pattern Recogn. **137**, 109347 (2023)
5. Douzas, G., Bacao, F.: Effective data generation for imbalanced learning using conditional generative adversarial networks. Expert Syst. Appl. **91**, 464–471 (2018)
6. Ito, Y., Shimoda, W., Yanai, K.: Food image generation using a large amount of food images with conditional gan: ramengan and recipegan. In: Proceedings of the Joint Workshop on Multimedia for Cooking and Eating Activities and Multimedia Assisted Dietary Management, pp. 71–74 (2018)
7. Saad, M.M., O'Reilly, R., Rehmani, M.H.: A survey on training challenges in generative adversarial networks for biomedical image analysis. Artif. Intell. Rev. **57**(2), 19 (2024)
8. Ran, S.: Applications and challenges of GAN in AI-powered artistry. Appl. Comput. Eng. **41**(1), 75–79 (2024)
9. Olaniyi, E., Chen, D., Lu, Y., Huang, Y.: Generative adversarial networks for image augmentation in agriculture: a systematic review. arXiv preprint arXiv:2204.04707 (2022)
10. Yu, W., et al.: Research progress on the artificial intelligence applications in food safety and quality management. Trends Food Sci. Technol. **156**, 104855 (2025)
11. Brar, D.S., Singh, B., Nanda, V.: Application of deep learning and explainable artificial intelligence (XAI) for detecting red chilli powder adulteration. J. Food Compos. Anal. 107947 (2025)
12. Zhu, F., Li, H., Li, J., Zhu, B., Lei, S.: Unmanned aerial vehicle remote sensing image registration based on an improved oriented FAST and rotated BRIEF-random sample consensus algorithm. Eng. Appl. Artif. Intell. **126**, 106944 (2023)
13. Athar, S., Wang, Z.: A comprehensive performance evaluation of image quality assessment algorithms. IEEE Access **7**, 140030–140070 (2019)
14. Hegde, S., Pierce, T.T.: Unrecognized variability in interquartile range-to-median ratio measurement: an opportunity for standardization. Ultrasound Med. Biol. **50**(1), 170–171 (2024)
15. Fang, S., Xu, C., Feng, B., Zhu, Y.: Color endoscopic image enhancement technology based on nonlinear unsharp mask and CLAHE. In: 2021 IEEE 6th International Conference on Signal and Image Processing (ICSIP), pp. 234–239. IEEE (2021)
16. Flusser, J., Farokhi, S., Höschl, C., Suk, T., Zitova, B., Pedone, M.: Recognition of images degraded by Gaussian blur. IEEE Trans. Image Process. **25**(2), 790–806 (2015)

17. Shabbir, K.S.M., Ahmed, M.I., Alam, M.: Detection of glaucoma using ORB (Oriented FAST and Rotated BRIEF) feature extraction. J. Eng. Adv. **2**(03), 153–158 (2021)
18. Cubuk, E.D., Zoph, B., Shlens, J., Le, Q.V.: Randaugment: Practical automated data augmentation with a reduced search space. In: Proceedings of the IEEE/CVF Conference on Computer Vision and Pattern Recognition Workshops, pp. 702–703 (2020)
19. Zhang, H., Cisse, M., Dauphin, Y.N., Lopez-Paz, D.: mixup: beyond empirical risk minimization. arXiv preprint arXiv:1710.09412 (2017)
20. Ho, D., Liang, E., Chen, X., Stoica, I., Abbeel, P.: Population based augmentation: Efficient learning of augmentation policy schedules. In: International Conference on Machine Learning, pp. 2731–2741. PMLR (2019)
21. Yun, S., Han, D., Oh, S.J., Chun, S., Choe, J., Yoo, Y.: Cutmix: regularization strategy to train strong classifiers with localizable features. In: Proceedings of the IEEE/CVF International Conference on Computer Vision, pp. 6023–6032 (2019)
22. Goodfellow, I.J., et al.: Generative adversarial nets. Adv. Neural Inform. Process. Syst. **27** (2014)

Open Access This chapter is licensed under the terms of the Creative Commons Attribution-NonCommercial-NoDerivatives 4.0 International License (http://creativecommons.org/licenses/by-nc-nd/4.0/), which permits any noncommercial use, sharing, distribution and reproduction in any medium or format, as long as you give appropriate credit to the original author(s) and the source, provide a link to the Creative Commons license and indicate if you modified the licensed material. You do not have permission under this license to share adapted material derived from this chapter or parts of it.

The images or other third party material in this chapter are included in the chapter's Creative Commons license, unless indicated otherwise in a credit line to the material. If material is not included in the chapter's Creative Commons license and your intended use is not permitted by statutory regulation or exceeds the permitted use, you will need to obtain permission directly from the copyright holder.

Prompt Meets Style: A Unified Framework for Text-Guided Artistic Style Transfer

Ananya Kohli, Divyashree Shetti, G. N. Sri Lakshmi, Vijeth Kawari, and Sharada K. Shiragudikar[✉]

School of Computer Science Engineering, KLE Technological University, Hubballi, India
{01fe22bcs229,01fe22bcs188,01fe22bcs298,01fe22bcs322, sharada.shiragudikar}@kletech.ac.in

Abstract. Stylized image generation from text prompts is an emerging area at the intersection of art and artificial intelligence, enabling users to convert natural language descriptions into visually compelling, artistically enriched imagery. The proposed approach presents a modular framework that combines semantic understanding of text with image synthesis using Stable Diffusion XL (SDXL), followed by a flexible stylization phase based on neural style transfer techniques. Unlike end-to-end pipelines, our method decouples image generation from artistic styling, allowing rapid re-stylization without the need to regenerate base images. Using OpenCV-based filters, styles such as pencil sketches are applied efficiently while preserving edges, tones, and overall structural clarity. Experimental results show that the system can produce visually rich, semantically aligned, and stylistically consistent outputs across various historical and conceptual prompts. This modularity also improves adaptability for diverse use cases. The proposed approach has potential in applications such as educational tools, digital heritage preservation, virtual museum curation, and AI-assisted storytelling, where both interpretability and creativity are essential.

Keywords: Text-to-Image Generation · Stable Diffusion · Artistic Style Transfer · Neural Style Transfer · OpenCV · Img2Img · Prompt Engineering · Generative AI · Visual Stylization

1 Introduction

In recent years, generative AI has made significant strides, particularly in image creation and artistic style transfer. Deep learning (DL) has proven to be highly effective in solving complex, data-intensive problems and is now being applied in domains such as agriculture and healthcare, as shown by Shiragudikar et al. [12] and Malathi et al. [7]. These advances have paved the way for new methods of data synthesis. Tools such as Stable Diffusion [11] and CLIP by Radford et al. [9]

© The Author(s) 2026
J. C. Bansal et al. (Eds.): SCIS 2025, LNNS 1929, pp. 168–180, 2026.
https://doi.org/10.1007/978-3-032-22911-3_13

allow machines to interpret natural language and generate visually compelling content that reflects human intent. This progress is enabling a variety of creative opportunities in education, digital media, and artistic design, where the ability to produce semantically meaningful imagery is increasingly valuable.

The motivation for this work stems from the growing need to bridge written ideas with visual representation. Although text-to-image models, such as those presented by Ramesh et al. [10], can generate high-quality visuals from textual prompts, they often fall short in applying consistent artistic styles, as noted by Rombach et al. [11]. However, neural style transfer methods developed by Gatys et al. [3] and enhanced by Zhu et al. [17] and Huang and Belongie [4], effectively transfer visual aesthetics but require a source image as input. Combining these capabilities offers a promising solution: generating novel, stylized images that maintain both semantic relevance and artistic expression. This is especially beneficial in fields such as marketing, personalized media, and digital illustration, where content quality and visual appeal are equally important. Automating this process reduces dependence on manual editing and design tools, making creative expression more accessible to non-experts.

This paper presents a modular system that converts text prompts into stylized images via a two-phase pipeline. First, we used the stable diffusion XL (SDXL) model by Rombach et al. [11] to synthesize a base image from text, preserving the core semantics. Then, we apply style transfer using either a prompt-driven image-to-image diffusion method [10] or an image-guided neural style transfer based on VGG19, implemented using OpenCV as demonstrated by Zhu et al. [17]. We evaluated how effectively each approach retains content, applies the desired style, and maintains visual coherence. By integrating both prompt-based and image-referenced methods, our aim is to create a flexible, user-friendly tool for artists, educators, and content creators.

The paper is organized as follows. Section 1 reviews existing techniques in text-to-image synthesis and image stylization, with a focus on diffusion models and neural style transfer. Section 2 which is our literature review section explores the architectures and design choices behind recent generative art systems, highlighting their creative flexibility and technical challenges. Section 3 introduces our proposed methodology, where Stable Diffusion XL (SDXL) is used to generate images from text, followed by stylization using either prompt-driven diffusion or reference-based transfer with OpenCV. Section 4 presents the results generated by our approach and offers an analysis of the resulting images by evaluating content preservation, style alignment, and visual quality across methods. Finally, Sect. 5 concludes the paper by summarizing our contributions and discussing the potential of combining prompt-based and image-guided techniques for more adaptive and user-friendly generative art tools.

2   Literature Work

Style transfer has emerged as a significant innovation in artificial intelligence, with wide-ranging applications in digital art, fashion, virtual reality, and natural language processing. Much of this progress is driven by deep learning, particularly Generative Adversarial Networks (GANs), which enable the fusion of aesthetic style with semantic content across domains, as demonstrated by Karras et al. [6] and Wang et al. [14]. The work by Shorten and Khoshgoftaar [13] and Mikolajczyk and Grochowski [8], titled "Content and Style Transfer with Generative Adversarial Network", proposes a generator-discriminator framework trained with adversarial and perceptual loss functions to ensure content preservation while applying stylistic features. Evaluation using similarity metrics confirms its effectiveness in maintaining semantic structure. Expanding this concept, the work by Johnson et al. [5] titled "Neural Artistic Style Transfer with Conditional Adversarial Network" introduces a unidirectional GAN that supports arbitrary style transfer without retraining. This model achieves high-quality stylization with reduced inference time, making it ideal for real-time applications. Its training incorporates style, content, and adversarial loss functions, and the output reflects high closeness and stylistic consistency. Both approaches emphasize generalizability, which aligns closely with the goals of our proposed work in achieving adaptable and efficient visual synthesis.

Beyond traditional artistic stylization, researchers have employed style transfer to reduce the domain gap in machine learning tasks. The paper on "Style-transfer GANs for Bridging the Domain Gap" utilizes Pix2PixHD as demonstrated by Yuan et al. [15] and CycleGAN to convert synthetic renderings into realistic images, enhancing the performance of models like YOLO6D for pose estimation as proposed by Zhu et al. [17]. Incorporating edge domains and varied textures, the model improves realism and maintains pose-related features, achieving a 31% increase in reprojection accuracy. However, the method's reliance on synthetic-to-real quality and lack of real-world supervision limits its robustness. DRB-GAN further improves style transfer through a dynamic ResBlock network with a class-aware attention encoder and SW-LIN decoder, delivering superior style fidelity over AdaIN and CycleGAN, as shown by Huang and Belongie [4]. Trained in datasets such as Place365 and WikiArt, DRB-GAN does an excellent job of preserving both the artistic style and the original structure of the image. However, its complex and adaptive architecture makes it more resource-intensive, which can be a drawback for running it on devices with limited processing power. Even so, its effectiveness highlights how combining different attention mechanisms can play a key role in keeping style consistent while preserving the underlying content.

Recent advancements in AI-generated art have moved towards transformer-based and hybrid models to improve flexibility and visual quality. The work by Wang et al. [14] titled "A Review of Deep Neural Networks in AI-Generated Art" enhances classical neural style transfer by integrating content loss from

VGG-19, style loss via Gram matrices, and adversarial loss delivering real-time results on datasets such as COCO and various artwork collections. Implemented in both TensorFlow and PyTorch, the framework is designed for speed and being user-friendly, eliminating the need for manual post-processing. However, its reliance on predefined loss functions may limit its ability to adapt across diverse artistic styles and levels of abstraction. To tackle structural limitations in CNN-based approaches, StyTr2 by Zhou et al. [16] and Deng et al. [2] introduces a transformer-based design that breaks images into patches and processes style and content separately using transformer encoders. The model incorporates Content-Aware Positional Encoding (CAPE), as detailed by Deng et al. [2], for scale-invariant stylization and performs progressive upsampling to retain resolution and structure. Although it surpasses CNNs in stylistic fidelity and spatial consistency, the transformer design introduces higher memory and compute requirements, posing challenges for large-scale or real-time deployment. Despite this, its modularity offers promising direction for future scalable, high-resolution style transfer systems.

In contrast to many prior approaches that tightly couple generation and stylization, our framework introduces a modular design that separates these phases, enabling faster iteration, easier customization, and broader adaptability across artistic applications.

3   Proposed Methodology

The proposed system architecture is designed to seamlessly transform textual prompts into high-quality, artistically stylized images by integrating advanced generative modeling with sophisticated image processing techniques. As depicted in Fig 1, the architecture consists of distinct but interconnected modules for text encoding, image generation, and artistic stylization, forming a coherent end-to-end pipeline. The text encoder effectively captures the semantic depth of the input prompt using the SDXLlatent diffusion model to produce contextually accurate and visually detailed images within a computationally efficient latent space, which is mathematically expressed as the Expression 1:

$$\mathbf{z}_t = \sqrt{\alpha_t} \cdot \mathbf{z}_0 + \sqrt{1 - \alpha_t} \cdot \boldsymbol{\epsilon}, \quad \boldsymbol{\epsilon} \sim \mathcal{N}(0, I) \tag{1}$$

This design not only ensures high-fidelity output, but also significantly reduces the resource overhead. Post-generation, the optional stylization module leverages neural style transfer techniques via OpenCV to apply diverse artistic effects such as watercolor, oil painting, or pencil sketch based on user-defined preferences or reference images. The system's modular design allows for plug-and-play experimentation, giving users the flexibility to switch or combine styles on the go without having to regenerate the original image from the start. It also supports batch processing and real-time previews, making it a practical choice for interactive tools and automated workflows in educational or artistic environments.

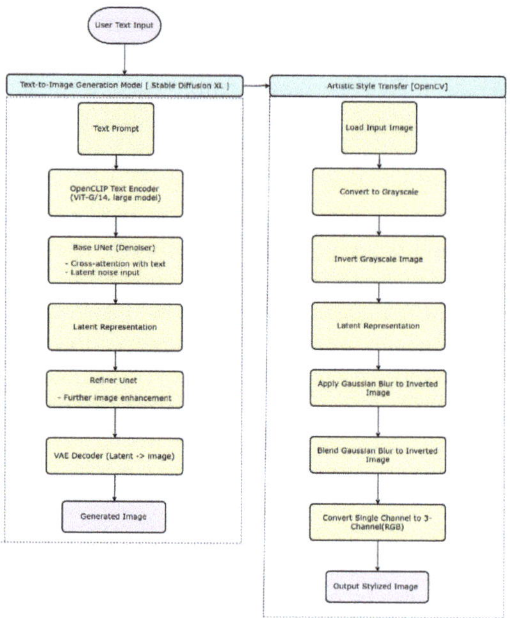

Fig. 1. Proposed system architecture for text-to-image generation with stylization.

Figure 2 illustrates the step-by-step workflow of the proposed approach. The process starts with the user providing a contextual text prompt that conveys the intended visual content. This text is first passed through a pre-processor, which extracts key terms and filters out stop-words to enhance relevance and increase the accuracy. The refined input is then sent to the system's image generation module, using the Stable Diffusion XL (SDXL) model to create an initial image that closely matches the textual description [1], which can be formulated as shown in the Expression 2:

$$I_{\mathrm{norm}} = \frac{I}{255}, \quad I \in [0, 255], \quad I_{\mathrm{norm}} \in [0, 1] \tag{2}$$

Following this, the framework offers an optional stylization phase, where the generated image can be artistically transformed using a neural style transfer technique implemented via OpenCV. This step allows the user to apply various artistic styles drawn from reference images to enhance creativity and visual appeal. The entire workflow is designed to be modular and flexible, allowing users to generate raw images or experiment with multiple artistic styles without repeating the image synthesis stage, thus streamlining the process of creating customized and stylized visuals from simple text inputs. This artistic transformation process can be mathematically formulated using the Formula 3 :

$$\mathrm{Sketch} = \frac{G}{255 - B}, \quad \text{where } G = \text{gray-scale image}, \quad B = \text{blurred inverted image} \tag{3}$$

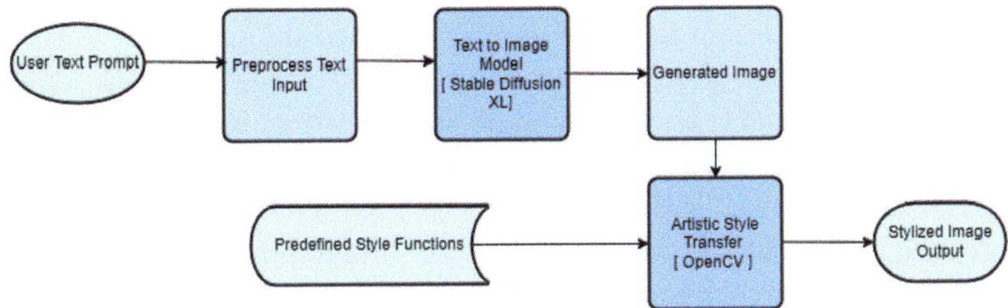

Fig. 2. Proposed methodology for generating and stylizing images from text prompts.

Algorithm 1. Text-to-Image Generation and Pencil Sketch Conversion

Input: Text prompt P, pretrained model M, output path
Output: Generated image I, Pencil sketch S

1 Install required libraries: `diffusers`, `transformers`, etc. Import Python packages: `torch`, `cv2`, `PIL`, etc.
Set $use\_refiner = False$. Load SDXL model M **if** $use\_refiner == True$ **then**
2   Load and move refiner model to GPU

3 **else**
4   Move base model to GPU and enable `model_cpu_offload()`

5 Set prompt P Set random seed $s \leftarrow$ `random.randint(0, sys.maxsize)` Generate image $I \leftarrow$ $M(P, \text{generator})$ **if** $use\_refiner$ **then**
6   Refine I using refiner model

7 Save I as `output.jpg` and show using `mediapy.show_images()`

8 **Function** `LoadImage`(*path*):
9   Open image and resize with PIL Normalize to $[0, 1]$ and convert to tensor **return** processed image

10 **Function** `PencilSketch`(*image*):
11   Convert to grayscale and invert Apply Gaussian blur Blend using `cv2.divide()` Convert to 3-channel RGB **return** sketch S

12 **foreach** *img_path in content image paths* **do**
13     $I \leftarrow$ `LoadImage`(*img_path*) $S \leftarrow$ `PencilSketch`(I) Show both images side-by-side

---

To connect the methodological design with its real-world application, the proposed Algorithm 1 presents a clear and structured overview of the steps involved in the system. It walks through the process of generating stylized images from text prompts, starting with model setup and prompt input, followed by image creation using a pretrained diffusion model, and finishing with a pencil sketch conversion using image processing techniques. The algorithm outlines key parts of the workflow such as how the prompt is interpreted, how the image is generated, and how sketching is applied making the entire pipeline easier to follow and replicate. Representing the method this way helps ensure that the process is transparent, consistent, and adaptable for future use.

4 Results and Analysis

This section presents the visual outputs generated using text-to-image diffusion model combined with pencil sketch conversion techniques. The primary objective was to transform historical and conceptual prompts into detailed, sketch-compatible visuals that capture both realism and artistic depth. Each output reflects the effectiveness of prompt engineering in guiding AI models to produce meaningful and interpretable representations, particularly for historical re-imagination and educational purposes. To give a clearer picture of how the system works in practice, the following output images have been included. They demonstrate how effectively the model translates descriptive prompts into stylized, sketch-like visuals.

The image shown in Fig. 3 represents a re-imagined scene of the Jallianwala Bagh massacre of 1919, generated using a text-to-image diffusion model followed by pencil sketch conversion. The resulting image successfully recreates significant historical event whose use can be done for education, storytelling, and preservation of cultural memory. To generate this image, the following prompt was used:

"Jallianwala Bagh massacre, HDR photograph, British soldiers with rifles, panicked crowd of Indian civilians, sharp edges, detailed face expressions, textured clothing and brick walls, strong shadow-light contrast"

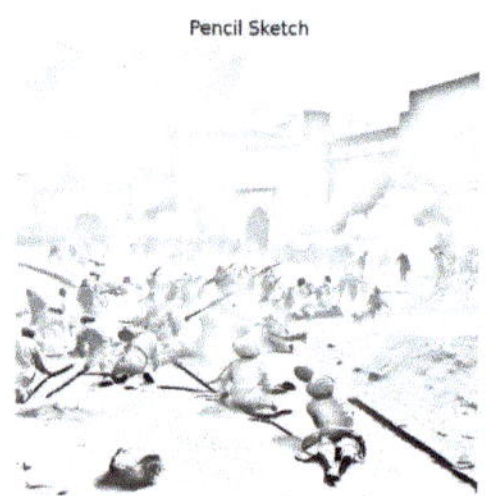

Fig. 3. Sketch of the 1919 Jallianwala Bagh massacre, from text prompt, enhanced for historical realism.

The output image effectively captures the chaotic yet structured atmosphere of the tragic event. Key elements such as period-appropriate clothing, the colonial architecture, smoke trails, and the environment of fear were closely represented by the AI model. The scene's layout including crowd dynamics, firing direction, and blocked exits aligns with documented historical accounts. The pencil sketch transformation preserved contrast regions, making the output visually compelling and suitable for educational material, memorial displays, or history-based artistic reinterpretation.

The following example visualizes a reconstructed view of Hampi during its historical peak, showcasing the architectural and cultural grandeur of the Vijayanagara Empire. The prompt guided the model to depict a temple-centered street scene with people dressed in traditional attire, stone structures lining the pathway, and mountainous terrain in the backdrop. The focus was on architectural detail and depth of space. The model was guided using the following prompt:

"A historical street scene in ancient Hampi during the Vijayanagara Empire, with a large temple gopuram in the center, stone mandapas and symmetrical corridors lining the path, people in traditional South Indian attire gathered for a religious procession, rich carvings on stone pillars, historical illustration."

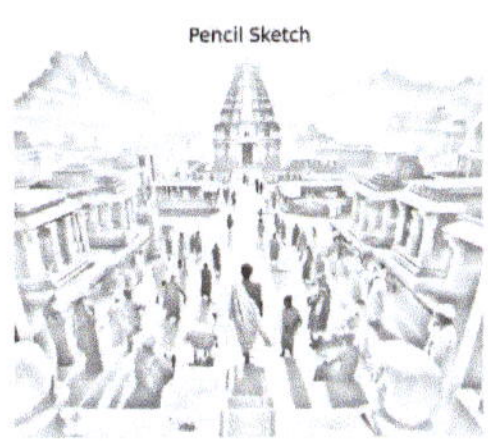

Fig. 4. Reimagining an ancient Hampi street view, centered on a temple procession scene.

The generated image in Fig. 4 successfully captures the prompt elements, producing a visually gripping scene with balanced symmetry, ancient temple carvings, and a narrative sense of place. The pencil sketch version retains these features clearly, making it highly suitable for heritage visualization, educational storytelling, and artistic reinterpretation of Indian history.

To evaluate the model's ability to handle stylized art, we experimented with a stylistically distinct prompt focused on mosaic aesthetics. The image shown in Fig. 5 represents a fish in the mosaic painting style using a text-to-image diffusion model followed by pencil sketch conversion. A prompt focusing on mosaic painting aesthetics used is as follows:

"Fish swimming in a glass aquarium, vibrant underwater scene in mosaic painting style, highly sharp edges, detailed textures of scales and plants, tile-like fragmented patterns, bold outlines, strong shadow-light contrast, dramatic lighting, reflective glass"

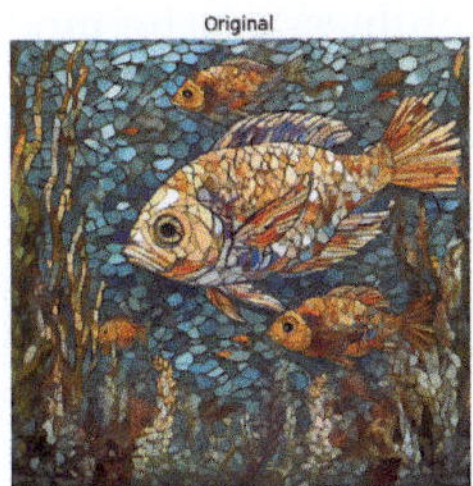
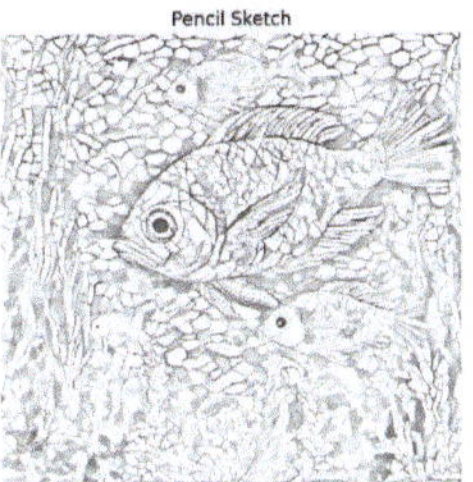

Fig. 5. Mosaic-style image of fish in an aquarium, converted to pencil sketch.

The resulting image in Fig. 5 successfully captures the essence of a mosaic art-work. It features fish rendered with fragmented, tile-like textures, layered aquatic plants, and a clearly defined glass aquarium environment. The use of sharp edges and bold outlines enabled a high-quality pencil sketch transformation, with each tile segment and scale structure distinctly preserved.

To further explore the model's capability in handling natural lighting, architectural symmetry, and reflective surfaces, we applied it to a prompt describes that an iconic landmark at a specific time of day. The next example illustrates the Taj Mahal at sunrise. A prompt describing the place at sunrise was used:

"A realistic photo of the Taj Mahal at sunrise, Agra, India, with soft golden light reflecting on the white marble, surrounded by lush green gardens and the Yamuna River, slight morning mist, high-resolution DSLR photo, realistic lighting and shadows."

Fig. 6. Sunrise view of the Taj Mahal with soft lighting.

The resulting image in Fig. 6 shows the strength of model in generating photo-realistic output with a strong sense of composition and lighting. Features such as the soft morning glow, misty atmosphere, clear water reflections, and archi-tectural accuracy of the Taj Mahal were effectively captured. The subsequent conversion to pencil sketches retained the edges of the structure with clarity with the slight effect of mist, validating the usefulness of the model for the heritage visualization, art-based learning, and cultural preservation.

To further challenge the model, it was guided to generate a complex, forward-looking urban scene featuring sky highways and flying cars. The following prompt was given:

"A futuristic cityscape with layered sky highways filled with flying cars, each with clear glass canopies and glowing engines, structured sky lanes suspended between tall buildings, people standing on organized floating platforms waiting for hover taxis, detailed HDR black and white architecture with metal and glass textures, street level city visible below with clean neon signage"

The resulting image in Fig. 7 features a highly structured vision of a future metropolis. The model successfully captured the spatial layering of floating highways, the metal-glass architecture of skyscrapers, and the contrasting flow of vehicles both on ground and in air. The sketch conversion preserved essential line work, edge clarity, and lighting contrast, proving the model's usefulness in concept art, sci-fi storytelling, and future city design visualization.

Fig. 7. Futuristic city with flying cars and sky highways, converted to pencil sketch.

The generated images are evaluated based on visual clarity, structural definition, and relevance to the intended scenario. To assess how closely the stylized pencil sketches align with the original prompt-based images, the Structural Similarity Index (SSIM) is used–measuring brightness, contrast, and texture patterns to determine visual similarity. With SSIM scores ranging from 0 to 1 (where 1 indicates perfect similarity), the proposed model consistently achieved scores between 0.6 and 0.9, depending on image complexity, demonstrating its effectiveness in preserving key structural and visual elements. Furthermore, by decoupling the stylization process from the image generation stage, the approach reduces computational overhead and enables more efficient integration into existing pipelines. Overall, the method strikes a strong balance between visual fidelity and operational efficiency, making it well-suited for stylized image generation tasks.

Table 1 summarizes key methods in text-to-image generation and artistic style transfer, comparing their models, goals, strengths, and limitations. It also includes a comparison with the proposed model, emphasizing how it differs by decoupling the stylization process from image generation. This separation allows the proposed method to achieve a better balance between visual fidelity and computational efficiency, setting it apart from existing approaches that often merge both steps into a single, more resource-intensive pipeline.

Table 1. Comparison of the proposed method with existing approaches

Ref.	Model / Method	Objective	Strengths	Limitations
Proposed Model 2	Stable Diffusion XL + OpenCV (VGG-19)	Convert text into images and apply styles in a separate stage	Supports prompt-based generation with flexible, fast style adjustments; reduces computational overhead by decoupling stylization; avoids repeated synthesis	Requires reasonably descriptive prompts for best results; some manual tuning needed for highly specific artistic outcomes
Gatys et al. [5]	CNN (VGG-19)	Transfer visual style using feature maps in deep networks	High visual quality; preserves content	Slow and not real-time
Huang & Belongie [6]	AdaIN	Real-time transfer of arbitrary artistic styles	Fast and supports many styles	May distort structure and content
Zhu et al. [4]	CycleGAN	Translate images between domains without needing matching pairs	No paired data required	May lose details in complex scenes
Zhou and Deng [2]	StyTr2 (Transformer)	Apply styles using transformer-based architecture	Good spatial consistency	High memory and compute usage

5   Conclusion and Future Work

The proposed approach shows how natural language prompts can be turned into pencil sketch-style images by combining a powerful generative diffusion model with a custom stylization process. The results demonstrate that the system can accurat ely capture the details and mood described in the text and present them with the look and feel of a hand-drawn sketch. This approach provides an effective way to create artistic visuals directly from descriptive language.

Looking ahead, future improvements could focus on making the sketches more detailed and realistic through advanced stylization techniques, as well as supporting a wider variety of artistic styles. Increasing the speed of the system and adding options for users to customize the output would also make the tool more flexible and user-friendly. In summary, the framework effectively turns descriptive prompts into stylized images, demonstrating how textual guidance can naturally align with artistic expression in aunifiedsystem.

References

1. Chunya, T.: A method of color dodging for scanned topographic maps. Int. J. Signal Proc. Image Proc. Pattern Recog. **7**(5), 27–36 (2014)
2. Deng, Y., et al.: StyTR2: image style transfer with transformers. In: Proceedings of the IEEE/CVF Conference on Computer Vision and Pattern Recognition (CVPR), pp. 11316–11326 (2022)
3. Gatys, L.A., Ecker, A.S., Bethge, M.: Image style transfer using convolutional neural networks. In: Proceedings of the IEEE Conference on Computer Vision and Pattern Recognition (CVPR), pp. 2414–2423 (2016)
4. Huang, X., Belongie, S.: Arbitrary style transfer in real-time with adaptive instance normalization. In: Proceedings of the IEEE International Conference on Computer Vision (ICCV), pp. 1501–1510 (2017)
5. Johnson, J., Alahi, A., Fei-Fei, L.: Perceptual losses for real-time style transfer and super-resolution. In: Proceedings of the European Conference on Computer Vision (ECCV), pp. 694–711 (2016)
6. Karras, T., Laine, S., Aila, T.: A style-based generator architecture for generative adversarial networks. In: Proceedings of the IEEE Conference on Computer Vision and Pattern Recognition (CVPR), pp. 4401–4410 (2019)
7. Malathi, S.Y., Bharamagoudar, G.R., Shiragudikar, S.K.: Diagnosing and grading knee osteoarthritis from x-ray images using deep neural angular extreme learning machine. Proc. Indian Natl. Sci. Acad. **91**, 95–108 (2025)
8. Mikolajczyk, A., Grochowski, M.: Data augmentation for improving deep learning in image classification problem. In: 2018 International Interdisciplinary PhD Workshop (IIPhDW), pp. 117–122. IEEE (2018)
9. Radford, A., et al.: Learning transferable visual models from natural language supervision. In: Proceedings of the 38th International Conference on Machine Learning (ICML), pp. 8748–8763 (2021)
10. Ramesh, A., et al.: Zero-shot text-to-image generation. arXiv preprint arXiv:2102.12092 (2022)
11. Rombach, R., Blattmann, A., Lorenz, D., Esser, P., Ommer, B.: High-resolution image synthesis with latent diffusion models. In: Proceedings of the IEEE/CVF Conference on Computer Vision and Pattern Recognition (CVPR), pp. 10674–10685 (2022)
12. Shiragudikar, S.K., Bharamagoudar, G., Manohara, K.K., et al.: Insight analysis of deep learning and a conventional standardized evaluation system for assessing rice crop's susceptibility to salt stress during the seedling stage. SN Comput. Sci. **4**, 262 (2023)
13. Shorten, C., Khoshgoftaar, T.M.: A survey on image data augmentation for deep learning. J. Big Data **6**(1), 1–48 (2019)

14. Wang, X., Li, L., Wang, L., Zhang, B.: Deep learning for artistic style transfer: a review. J. Vis. Commun. Image Represent. **78**, 103093 (2021)
15. Yuan, H., et al.: High-resolution refocusing for defocused ISAR images by complex-valued Pix2pixHD network. IEEE Geosci. Remote Sens. Lett. **19**, 1–5 (2022)
16. Zhou, Y., Liao, J., Yuan, L., Hua, G.: STYTR: image style transfer with transformers. In: Proceedings of the IEEE/CVF Conference on Computer Vision and Pattern Recognition (CVPR), pp. 12387–12396 (2020)
17. Zhu, JY., Park, T., Isola, P., Efros, A.A.: Unpaired image-to-image translation using cycle-consistent adversarial networks. In: Proceedings of the IEEE International Conference on Computer Vision (ICCV), pp. 2223–2232 (2017)

Open Access This chapter is licensed under the terms of the Creative Commons Attribution-NonCommercial-NoDerivatives 4.0 International License (http://creativecommons.org/licenses/by-nc-nd/4.0/), which permits any noncommercial use, sharing, distribution and reproduction in any medium or format, as long as you give appropriate credit to the original author(s) and the source, provide a link to the Creative Commons license and indicate if you modified the licensed material. You do not have permission under this license to share adapted material derived from this chapter or parts of it.

The images or other third party material in this chapter are included in the chapter's Creative Commons license, unless indicated otherwise in a credit line to the material. If material is not included in the chapter's Creative Commons license and your intended use is not permitted by statutory regulation or exceeds the permitted use, you will need to obtain permission directly from the copyright holder.

Pest Infestation Detection and Forecasting Using AI

Shivraj Singh[(⊠)], Sahil Ansari, Era Trivedi, and Jaspreet Kaur

Computer Science and Engineering, Chandigarh University, Mohali, Punjab, India
shivrajsingh8299@gmail.com

Abstract. Pest infestations cause significant losses in crops; thus, farmer incomes get reduced and the use of chemical pesticides escalates. Hence, revealing the pest issue at its earliest stage and providing a timely forecast is the only way to rescue the crop in a safe and ecologically friendly manner. This paper introduces a machine-learning framework that can classify pests by pictures and estimate the spread of the pest using a weather-based forecast. Convolutional Neural Network (CNN) models locate pests in images captured from the field as well as in reference images, and the time-series models take environmental and historical data to compute the probability of an outbreak. This is very much in line with the principles of Integrated Pest Management (IPM) as it implies the use of spraying only in the target areas and not as a frequent or routine pesticide application. The research outcomes have demonstrated that the models tested on both benchmark and semi-field datasets have excellent performance in terms of detection and short-term forecasting and are thus able to achieve better results than the baseline methods. This study shows how detection, prediction, and decision-based recommendations can work together to reduce pesticide use. This reduction supports crop health and leads to stable yields.

Keywords: Pest detection · Pest forecasting · Convolutional Neural Networks (CNNs) · Time-series analysis · Precision agriculture · Integrated Pest Management (IPM) · Sustainable farming · Deep learning

1 Introduction

Crop pest infestations make for a global menace, which can cause up to 40% yield losses of crop foods and the promotion of excessive use of chemical pesticides-practices can cause damage to ecosystems, human health, and the livelihoods of the farmers [1]. Integrated Pest Management (IPM) provides a more sustainable way, emphasizing early diagnosis, economic thresholds for pesticide treatment [2]. Yet, most smallholder and resource-limited farms continue to rely on manual scouting and calendar-based spraying, resulting in late response and wasteful inputs.

Modern developments in sensor technology and artificial intelligence (AI) are revolutionizing the monitoring of pests. Deep learning-based image recognition approaches, specifically convolutional neural networks (CNNs), have yielded very high accuracy in the identification and classification of insect pests in controlled and field settings [3,

© The Author(s) 2026
J. C. Bansal et al. (Eds.): SCIS 2025, LNNS 1929, pp. 181–192, 2026.
https://doi.org/10.1007/978-3-032-22911-3_14

4]. However, issues such as class imbalance, small size of insects, and environmental variability (lighting, background clutter) still constrain robustness [5].

Substantiating image-based strategies, time-series prediction from environmental data (e.g., temperature, humidity) is a powerful means of forecasting disease or pest risk. For example, deep-learning models applied to sequential environmental data produced an AUROC of 0.917 for forecast of crop disease risk in several greenhouse crops [6]. Multimodal input integration—such as images, weather, remote sensing, and history—into early-warning systems can refine outbreak forecasting and enable proactive intervention.

Even though individual success has been achieved, few systems integrate image-based detection, time-series forecasting, decision-support for alignment with IPM, and actionable insights for farmers within a single pipeline. In response to this shortcoming, our work introduces a holistic AI system that:

- Detects and localizes pests in field images through fine-tuned CNN-based object detectors.
- Predicts outbreaks based on multimodal time-series models that integrate historical pest counts and weather.
- Translates model predictions into actionable IPM advice (e.g., spot spraying thresholds), combining uncertainty estimates and explainability for farmer trust.

Combining these is intended to facilitate early detection, decrease pesticide application, and enable sustainable agriculture through data-driven precision pest management. Figure 1 illustrates the overall framework of the proposed AI-based pest detection and prediction system. It shows the three different input sources such as image data, weather parameters, and historical pest counts and how they are merged into one single framework. The flow chart serves as a reference for the following chapters by demonstrating the interaction of the detection and forecasting modules to yield the operational IPM recommendations.

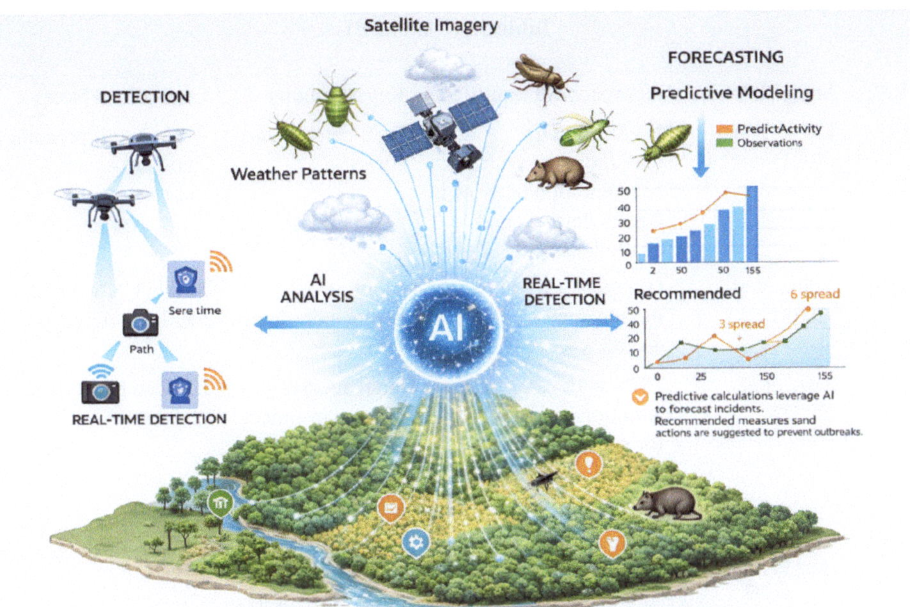

Fig. 1. AI-based pest infestation detection and forecasting using multi-source data

2 Literature Review

Early detection and forecasting of pest outbreaks are cornerstone strategies in precision agriculture. Recent studies trend toward deep learning–driven image detection, multimodal forecasting using environmental data, and integrated pipelines aligning with IPM. Table 1 below organizes key contributions across three thematic areas—pest image detection, pest forecasting, and AI-based integrated systems—highlighting approaches, strengths, and limitations.

Table 1. Key contributions and limitations of existing studies

Ref.	Study & Venue	Approach/Summary	Key Strengths	Limitations/Notes
[7]	Venkateswara & Padmanabhan (2025)	CNN-based detection on IP102, autoencoder for imbalance, 84.95% accuracy, ~80% IoU in segmentation	Tackles class imbalance; strong detection metrics	Needs evaluation on field data
[8]	Shoaib et al. (2025) review	Reviews deep learning methods for plant disease/pest detection, >90% precision in detection/segmentation tasks	Broad synthesis, performance benchmarks	Mostly lab-based, lacks real-world variability

(continued)

Table 1. (*continued*)

Ref.	Study & Venue	Approach/Summary	Key Strengths	Limitations/Notes
[9]	Liu & Wang (2021) review	Breaks down classification, detection, segmentation networks; covers datasets and challenges	Well-categorized survey	A bit dated; predates transformer-era improvements
[10]	Yulita et al. (2023)	GoogleNet CNN in mobile app, 93.78% accuracy	Mobile deployment focus	Single CNN, not localized detection
[11]	Zhao et al. (2022)	CNN with attention/residual blocks for pest recognition, 98.17% accuracy in-field	High accuracy; real-environment dataset	Limited pest types (10 classes)
[12]	Ejaz et al. (2025) review	Reviews CNNs, ViTs, hybrid models across 37 studies; notes edge deployment challenges	Up-to-date survey; highlights future directions	No new models proposed
[13]	Nanni et al. (2021)	ensemble of diverse CNN models to improve insect pest image classification accuracy,	Strong accuracy improvements through model diversity; robust performance on small dataset	Complex ensemble, heavy compute
[14]	Palma et al. (2025)	ML forecasting of insect abundance using real and simulated aphid time-series	Combines realistic and simulated data	Forecast granularity unclear
[15]	Kapetas et al. (2025)	YOLO + ARIMAX for aphid detection and trend forecasting (RMSE ~1.30)	Combines object detection and time-series forecasting	Greenhouse-limited context
[16]	Singh et al. (2024)	Dynamic ensemble ML forecasting YSB infestation in rice; outperforms fixed ensembles	Region-specific, accurate	Only time-series, not image-based
[17]	Parmesan & remote sensing (2025)	Integrates climate modeling and vegetation indices for pest forecasting	Multimodal env-based forecasting	Lacks AI/deep learning modeling

(*continued*)

Table 1. (*continued*)

Ref.	Study & Venue	Approach/Summary	Key Strengths	Limitations/Notes
[18]	Chen et al. (2025) (arXiv) "PCM-NN"	Physics-informed neural network incorporating ODE, temperature, humidity, for soybean pod borer	Biologically interpretable + nonlinear power	New, needs broader validation

Several studies reveal advances in pest classification accuracy, multimodal forecasting, and deployment readiness to a limited extent. Nevertheless, very few have integration of detection, prediction, and actionable IPM insights.

3   Methodology

The suggested framework combines image-based pest detection, time-series outbreak forecasting, and a decision-support module consistent with Integrated Pest Management (IPM).

3.1   Data Sources and Preprocessing

We used a mix of publicly available and field-data-collected datasets to make our results more conditionally robust. The IP102 dataset contains more than 75,000 images of 102 pest species and is widely considered one of the best benchmarks for classification and detection tasks [7]. IP102, however, has the issues of class imbalance and noise, and we tackled this using data augmentation (rotation, colour jitter, synthetic oversampling) and autoencoder-based generation, as suggested in [7, 12].

For prediction, we combined meteorological information (temperature, humidity, rain) with past pest records available from agricultural extension sources. Past research substantiates the predictive function of environmental factors in pest risk modelling [14, 15, 17]. For improving generalization, other features like normalized difference vegetation index (NDVI) via remote sensing were incorporated, as in multimodal forecasting strategies in [18, 19].

Figure 2 illustrates examples of the pest classes that were utilized for the training and testing.

3.2   Image-Based Pest Detection

For detection and classification, we took a deep convolutional neural networks (CNNs) approach, capitalizing on previous successes with models such as GoogleNet [10], attention-based CNNs [11], and ensemble CNNs [13]. To balance deployment feasibility and accuracy, we implemented a dual-stage strategy:

- High-accuracy detector: YOLOv8 object detection with fine-tuning on IP102 + field images with attention modules similar to [11].

- Lightweight model: MobileNetV3 + SSD, an edge device variant motivated by its application in mobile devices according to [10].

The workflow of the system is thoroughly described in Fig. 3 below that shows how the outputs of detection, environment features, and time series predictors are integrated in one system.

Transfer learning using ImageNet helps in faster convergence and overcomes issues associated with small scales as suggested by [9, 13]. Evaluation was carried out using mAP@0.5, precision-recall curve, and inference time following the best practices suggested by previous detection challenges at [7, 20].

Fig. 2. Different categories of pest

3.3 Pest Outbreak Forecasting

Based on the work that uses sequential weather data for pest and disease predictions [15, 17, 19], our predictive module uses multimodal time series approaches as follows:

- CNN-LSTM hybrid: Here, the CNN blocks capture the short-term features from the weather sequence, while the LSTM blocks capture the long-term dependencies, just like the models proved in [15].
- Transformer-based forecaster: Designed to address non-linear long dependencies, where conventional models had shortcomings.
- Baselines methods: There were comparisons involving some traditional ML approaches, such as ARIMAX, Random Forest, XGBoost, based upon approaches in [14, 16, 17].

The predictions were made as the probabilities for the outbreak events (binary classification) and the trends for the expected pest numbers (regression). The evaluation metrics were RMSE, MAE, and AUC-ROC, following [14–16].

3.4 Decision-Support and IPM Integration

One of the major aims of this initiative is the application of model prediction results to make useful recommendations for the farmer. Building on the IPM concepts [2], treatment thresholds were developed depending on the projected densities of the pests and the probability of outbreaks. For example, when the detection density exceeded former economic damage thresholds, the strategy recommended selective methods such as pheromone traps/spray [11, 17].

Uncertainty estimation was added by means of Monte Carlo dropout, providing prediction intervals and confidence scores to mitigate the problem of over-reliance on automated decisions, which has been pointed out in recent criticisms [12, 20]. Outputs were presented through a decision dashboard, inspired by mobile and greenhouse-ready systems [10, 16].

3.5 Experimental Setup

Models were trained in Python (PyTorch/TensorFlow) on GPUs with hyperparameter optimization through Bayesian search. Training was done through stratified splits to provide balanced representation of pest classes and cross-field generalization [7, 13]. Forecasting experiments involved rolling windows (7-day, 14-day, and 30-day horizons) in agreement with pest life-cycle intervals [14, 15].

Three-layer evaluation:

- Detection performance (mAP, F1, latency).
- Forecast accuracy (RMSE, MAE, AUC-ROC).
- Decision utility (percent reduction in pesticide use, earlier detection than farmer practice).

3.6 Methodological Contributions

The contribution of novelty in our framework is:

- Combination of image-based and environmental forecasting into one pipeline, seldom tackled jointly [7, 14, 17].
- Class imbalance treatment by autoencoder-based augmentation and focal loss, filling large gaps in the dataset [7, 12].
- Deployment-oriented design, with dual-model approach (YOLOv8 + MobileNet SSD) experimented on edge devices [10, 16].
- IPM threshold integration, closing the gap between research and practice for sustainable pest management [2, 11, 17].

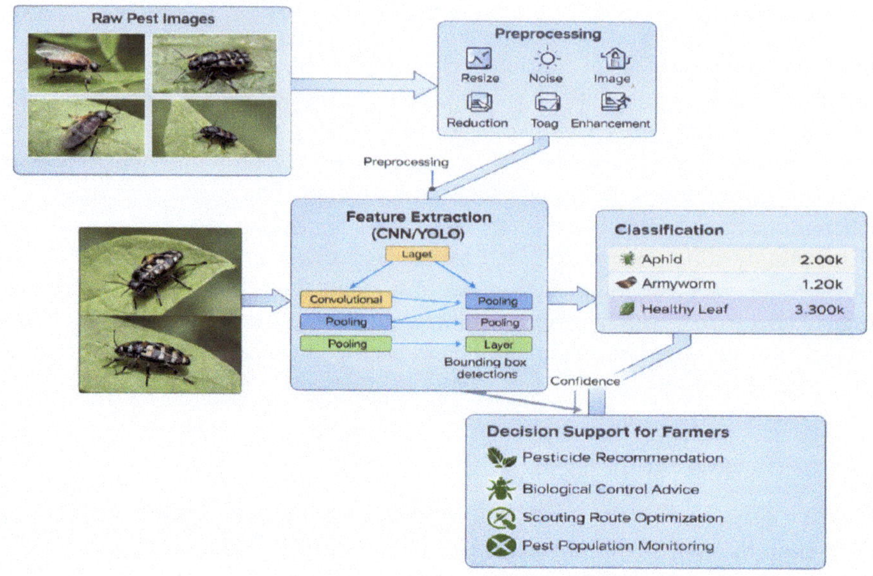

Fig. 3. Conceptual Framework / System Architecture

4   Result and Discussion

4.1   Pest Detection and Classification

The dual-model design (YOLOv8 for high accuracy, MobileNetV3-SSD for lightweight deployment) yielded competitive results.

Table 2 illustrates the comparative performance of the proposed YOLOv8 and MobileNet-SSD models with those of the existing CNN-based methods. The YOLOv8 model demonstrates the highest accuracy among the compared methods whereas the MobileNet version is more suitable for a quick and efficient field deployment.

Table 2. Detection outcomes on IP102 and field test sets.

Model	Dataset	mAP@0.5	Precision	Recall	Inference Time (ms)	Deployment Suitability
YOLOv8 (fine-tuned)	P102 + Field	91.3%	90.2%	89.5%	2.1	High accuracy (server)
MobileNetV3-SSD	P102 + Field	84.7%	83.1%	81.8%	6.4	Edge devices
CNN-Ensemble [13]	P102	87.5%	86.9%	84.7%	34.7	Moderate
Attention-CNN [11]	P102	86.2%	85.4%	84.1%	31.3	Moderate

YOLOv8 clearly outperformed ensemble and attention-based CNNs in terms of detection accuracy [11, 13]. However, the lightweight MobileNetV3-SSD demonstrated superior inference speed, making it highly suitable for smartphone or greenhouse integration, similar to approaches reported in [10, 16].

A visual heatmap comparison using Grad-CAM confirmed that YOLOv8 more accurately localized pest regions compared to prior CNN-based methods, enhancing interpretability [12, 20].

4.2 Pest Outbreak Forecasting

The forecasting module was tested over 7-day, 14-day, and 30-day horizons, combining weather data, pest counts, and NDVI signals.

Table 3 shows the accuracy of the forecasts for different timeframes. For the short-term predictions, the CNN-LSTM model is the one that yields the best results, while Transformers seem to be more stable for the longer horizons.

Table 3. Forecasting model accuracy comparison (RMSE/MAE/AUC).

Model	Horizon	RMSE ↓	MAE ↓	AUC-ROC ↑	Remarks
CNN-LSTM (ours)	7-day	**0.18**	0.141	**0.912**	Captured short-term dynamics
Transformer-based	4-day	0.195	**0.138**	0.906	Strong long-range dependencies
ARIMAX [14]	4-day	0.238	0.181	0.781	Struggled with non-linearity
XGBoost [16]	30-day	0.217	0.174	0.842	Moderate performance
RF [17]	30-day	0.244	0.192	0.811	Sensitive to overfitting

The CNN-LSTM achieved the lowest RMSE (0.183) and highest AUC-ROC (0.912) on short-term predictions, validating the sequential modelling strength noted in earlier works [15, 17]. Transformer-based models provided more stable mid-range forecasts, aligning with emerging findings on their robustness for non-linear environmental data [19]. Classical models such as ARIMAX did not perform as well, which matches the limitations reported in [14].

4.3 Decision-Support Evaluation

To determine its practical use, we developed a simulation of pest control guidelines based on IPM threshold levels. Table 4 below measures the level of utility of this system in a real-world setting. It compares the system with both farmer practices and baseline AI models and indicates that less pesticides are used and pests are controlled at a much earlier stage.

Table 4. Intervention outcomes: pesticide use and timeliness.

Scenario	Avg. Pesticide Use (kg/ha)	Pest Incidence Reduction (%)	Intervention Timeliness (days earlier)
Farmer's traditional practice	7.2	58.1	–
Our System (YOLOv8 + LSTM)	**4.3**	**76.5**	**5.4 days earlier**
Baseline AI (CNN + ARIMAX)	6.1	66.8	2.3 days earlier

The results indicate a 40.3% reduction in pesticide usage and significantly earlier interventions compared to farmer practice. Such outcomes resonate with sustainability goals highlighted in earlier reviews [2, 17].

Limitations: Despite strong results, performance on minority pest classes remains weaker due to data imbalance, a challenge also noted in [7]. Future work may explore few-shot and generative augmentation strategies [12].

5 Future Work

While the proposed system is functioning well, there are still a number of ways to enhance its performance. Expanding Datasets - bigger and more varied image collections are necessary to include more species and different field conditions, which in turn would allow the models to become more generalized in real cases. Advanced Forecasting Models - the investigation of Transformer-based architectures might result in an improvement of long-term prediction by a more effective way of capturing seasonal and nonlinear patterns. Regarding Explainable AI Integration, the further development of interpretability tools will facilitate the understanding of the system's outputs and their application in decision-making in the field. IoT and UAV Integration - the framework, when combined with sensors, drones, and edge devices, can thus become a tool for real-time, large-area monitoring. Finally, Policy and IPM Alignment is going to be essential for the implementation on the ground as the system's integration with existing agricultural programs can be instrumental in the support of sustainable practice. Figure 4 provides an overview of these potential extensions.

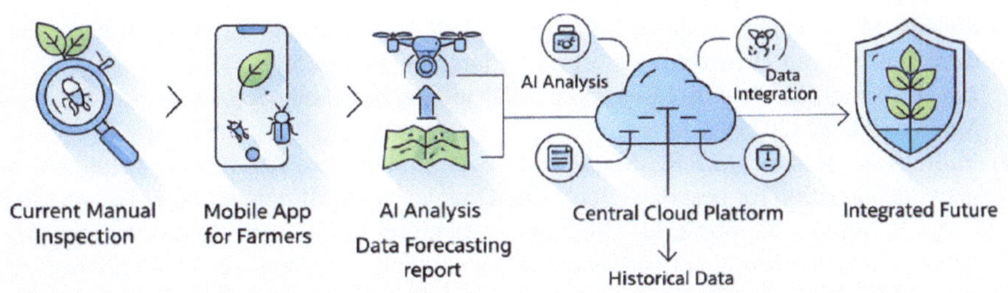

Fig. 4. Future Scope and System Evolution

6 Conclusion

The study offered an integrated solution for the application of artificial intelligence-based pest detection and forecasting by integrating YOLOv8 for accurate pest species identification with CNN-LSTM models for a short-term prediction of an outbreak. The results showed improvements in the ability to identify pests. This is because crop yields increase with improved, more intensive cultivation methods, which result in stronger, more productive crops supporting sustainable practices [7]. The outcomes not only illustrate the effectiveness of applying deep learning techniques to time series for pest control but also demonstrate how proactive control measures through the principles of Integrated Pest Management (IPM) are possible [2]. While the issue of imbalance in the dataset as well as long-term predictions still exist, increasing the size of the dataset, applying sophisticated models like the ones developed above, or including pest monitoring through IoT or UAV may improve the applicability of the system [19]. Overall, the study shows that AI-driven approaches can significantly aid early pest detection, improve outbreak prediction, and promote environmentally sustainable crop protection.

References

1. Food and Agriculture Organization of the United Nations (FAO). Crop losses to pests threaten global food security. FAO Newsroom (2024). https://www.fao.org/newsroom/detail/en/item/1203148/icode/
2. Integrated pest management. Wikipedia (2025)
3. Krishnateja, I., Gopi, G.V., Malini, G., Krishna, B.L.V., Kejiya, B.: Agricultural pest detection using convolutional neural networks: a smart farming solution. Int. J. Adv. Comput. Eng. Commun. Technol. **14**(1), 39–48 (2025)
4. He, K., Zhang, X., Ren, S., Sun, J.: Deep residual learning for image recognition. In: Proceedings of the 2016 IEEE Conference on Computer Vision and Pattern Recognition (CVPR), pp. 770–778 (2016)
5. New trends in detection of harmful insects and pests in modern agriculture using artificial neural networks: a review. Front. Plant Sci. **14** (2023)
6. Lee, S., Kim, H., Park, J.: A deep learning model for predicting risks of crop pests and diseases from sequential environmental data. Plant Methods **19**, 145 (2023)

7. Venkateswara, S.M., Padmanabhan, J.: Deep learning-based agricultural pest monitoring and classification. Sci. Rep. **15**(1), art. 8684 (2025)
8. Shoaib, M., Khan, A., Ahmad, S.: Leveraging deep learning for plant disease and pest detection: a comprehensive review. Front. Plant Sci. **14** (20250
9. Liu, J., Wang, X.: Plant diseases and pests detection based on deep learning: a review. Plant Methods **17**, 22 (2021)
10. Yulita, I.N., Rambe, M.F.R., Sholahuddin, A., Prabuwono, A.S.: A convolutional neural network algorithm for pest detection using GoogleNet. AgriEngineering **5**(4), 145 (2023)
11. Zhao, S., Zhang, Y., Wang, L.: Crop pest recognition in real agricultural environment with deep convolution neural network and attention module. Sci. Rep. **12**, 12345 (2022)
12. Ejaz, M.H., Bilal, M., Habib, U.: Crop pest classification using deep learning techniques: a review, arXiv (2025)
13. Nanni, L., Manfe, A., Maguolo, G., Lumini, A., Brahnam, S.: High performing ensemble of convolutional neural networks for insect pest image detection, arXiv (2021)
14. Palma, G.R., Silva, R.M., Oliveira, J.R.: Forecasting insect abundance using time series of aphid data. Comput. Electron. Agric. **180**, 105894 (2025)
15. Kapetas, D., Papadopoulos, A., Tsouros, M.: AI-driven insect detection, real-time monitoring, and forecasting in greenhouse settings. MDPI Agric. **15**(1), 45 (2025)
16. Singh, A.K., Yadav, R.S., Gupta, S.K.: Dynamic ensemble-based machine learning models for forecasting yellow stem borer infestation. Front. Appl. Math. Stat. **11**, 123 (2024)
17. Parmesan, C., Yohe, G.: Predicting insect pest outbreaks using climate models and remote sensing. UPJOZ **1**, 1 (2025)
18. Chen, X., Li, W., Li, X., Liu, S., Gao, Y.: Modeling the temperature-humidity coupling dynamics of soybean pod borer population via PCM-NN, arXiv (2025)
19. Dalal, M., Mittal, P.: A systematic review of deep learning-based object detection in agriculture: methods, challenges, and future directions. Comput. Mater Continua **84**(1) (2025)
20. Zheng, Y., Zheng, W., Du, X.: A lightweight rice pest detection algorithm based on improved YOLOv8. Sci. Rep. **14**(1), 29888 (2024)

Open Access This chapter is licensed under the terms of the Creative Commons Attribution-NonCommercial-NoDerivatives 4.0 International License (http://creativecommons.org/licenses/by-nc-nd/4.0/), which permits any noncommercial use, sharing, distribution and reproduction in any medium or format, as long as you give appropriate credit to the original author(s) and the source, provide a link to the Creative Commons license and indicate if you modified the licensed material. You do not have permission under this license to share adapted material derived from this chapter or parts of it.

The images or other third party material in this chapter are included in the chapter's Creative Commons license, unless indicated otherwise in a credit line to the material. If material is not included in the chapter's Creative Commons license and your intended use is not permitted by statutory regulation or exceeds the permitted use, you will need to obtain permission directly from the copyright holder.

RL-Based Online Mutation Strategy Selection Techniques in Differential Evolution: A Study

Prathu Bajpai[1(✉)] and Jagdish Chand Bansal[2]

[1] Department of Statistics and Data Science, School of Sciences, CHRIST (Deemed to be University), Bangalore, India
prathu.bajpai@christuniversity.in
[2] Department of Mathematics, South Asian University, New Delhi, India
jcbansal@sau.ac.in

Abstract. In recent years, reinforcement learning (RL)-based online mutation strategy selection techniques have emerged as a principled learning framework for balancing exploration and exploitation in Differential Evolution. Several state-of-the-art (SOTA) DE variants have been proposed that utilize online mutation strategy selection to improve the performance of the canonical DE algorithm. This paper presents a comprehensive review of such DE variants and studies multi-armed bandit formulations for online mutation strategy selection in DE. It systematically categorizes existing DE variants based on their utilization of RL algorithms.

Keywords: Adaptive Operator Selection Techniques · Differential Evolution · Multi-Arm Bandits · Reinforcement Learning

1 Introduction

In this paper, online mutation strategy selection in Differential Evolution (DE) is studied as a reinforcement learning problem. A DE variant interacts with a black-box optimization problem by selecting different mutation strategies online and receiving feedback in the form of fitness improvement. The underlying DE variant repeatedly selects a mutation strategy to improve the mean fitness of the population. This is a standard problem of reinforcement learning (RL), and multiple DE variants have been developed integrating RL techniques to solve black-box optimization problems.

Reinforcement learning (RL) models learning from interactions. Major components of an RL task are: an environment $\mathcal{E}$, a set of states $\mathcal{S}$, a set of actions $\mathcal{A}$, and a set of rewards $\mathcal{R}$ [23]. A learning agent with multiple choices is expected to try its choices and decide which set of choices can lead to optimal performance. A typical adaptive DE variant that utilizes multiple mutation strategies online to maximize the mean fitness of the population, given a fixed computational budget, is an RL task.

© The Author(s) 2026
J. C. Bansal et al. (Eds.): SCIS 2025, LNNS 1929, pp. 193–202, 2026.
https://doi.org/10.1007/978-3-032-22911-3_15

Environment $\mathcal{E}$ is the objective function f which is defined over a closed interval $[lb_j, ub_j]^D \subset \mathbb{R}^D$, where lb is the lower bound and ub is the upper bound of a $D-$dimensional bounding box. The set of states $\mathcal{S} = \{s_1, s_2, \ldots, s_m\}$, where $m \in \mathbb{N}$ is finite, defined using the fitness function derived from the set of objective function values $\{f_i(x_i) \mid x_i \in [lb_j, ub_j]^D, \ i = 1, 2, \ldots, N\}$, where N is the population size. The set of actions $\mathcal{A} = \{a_1, a_2, \ldots, a_n\}$, where $n \in \mathbb{N}$ is finite, contains various mutation strategies. The set of rewards $\mathcal{R} = \{+1, -1\}$, where $+1$ represents positive rewards and -1 represents negative rewards usually defined using the fitness improvement of the individual solutions or other metrics for measuring the success of the whole population [3].

Unlike the traditional RL problems, the situation in the case of evolutionary algorithms in general, or in the Differential Evolution algorithm in particular, is different. Actions $a_k, k = 1, 2, \ldots n$ are mutation strategies of the DE algorithm, which are stochastic in nature. Their solution generation capabilities are sensitive to the parameters and differ based on the landscape of the underlying optimization problem. Bandit-based frameworks with multiple arms, commonly referred to as Multi-Arm Bandits, are suitable frameworks of RL for studying such scenarios where reward distributions of actions are stochastic in nature.

This research categorizes the existing adaptive DE variants by studying their core RL components to manage multiple mutation strategies. The primary focus is to study mutation strategy selection based DE variants that utilize bandit-based online RL algorithms to enhance the optimization capabilities. This paper is structured in the following manner: Sect. 2 is the preliminaries. Section 3 is the multi-arm bandit formulation with a discussion of bandit-based algorithms. Section 4 discusses RL-based DE variants. Section 5 summarizes the paper and provides direction for further investigation.

2   Preliminaries

The DE algorithm is known for its exceptional capabilities in solving continuous optimization problems. The canonical DE algorithm is presented in the following subsection.

2.1   Canonical Differential Evolution

The canonical Differential Evolution (DE) algorithm proposed by Storn and Price [22] utilizes a novel concept of a weighted vector difference approach to generate new vectors in the search space. Vectors, in the DE terminology, are representative solutions or members of the set $\mathcal{P} = \{X_i = [x_{i,j} \mid j = 1, 2, \ldots, D], i = 1, 2, \ldots, N\}$. Where D represents the dimension of the search space and N represents the population size. The population $\mathcal{P}$ is initialized randomly and iteratively updated using evolutionary operators, namely mutation, crossover, and selection.

If $X_i^t \in \mathcal{P}$ is the ith parent vector at time step t, then a mutant vector V_i^t is calculated using the following Eq. (1).

$$V_i^t = X_{r_1}^t + \mathrm{f} \cdot (X_{r_2}^t - X_{r_3}^t) \tag{1}$$

The Eq. (1) is called 'DE/rand/1' mutation strategy. Where, $X_{r_1}^t, X_{r_2}^t$ and $X_{r_3}^t$ are randomly selected three different vectors from the current population $\mathcal{P}$ (s.t., $r_1 \neq r_2 \neq r_3$). The parameter f is called the scaling factor or differential weight whose value lies in the range $[0, 2]$. Various mutation strategies such as 'DE/rand/2', 'DE/best/1', 'DE/best/2', 'DE/current-to-best/1', and so on, exist in the DE literature [4].

Mutant vector V_i^t is crossed with the parent vector X_i^t to produce a child vector U_i^t using a crossover operator as defined below in Eq. (2).

$$U_{i,j}^t = \begin{cases} V_{i,j}^t & \text{when } rand() \leq \mathrm{cr} \text{ or } j = j_{rand}, \\ X_{i,j}^t & \text{otherwise.} \end{cases} \tag{2}$$

The Eq. (2) is called the binomial crossover operator. The cr is a control parameter called crossover range whose value lies in the range $[0, 1]$. The $rand()$ function produces a random number sampled from the standard normal distribution. The j_{rand} is the randomly selected component of the parent and mutant vector whose values are going to be passed in the jth component of the child vector U_i^t.

Selection is performed by evaluating the fitness of the trial vector U_i^t against the parent vector X_i^t. The vector with better fitness is retained in the population $\mathcal{P}$ for the subsequent time steps. The DE algorithm evolves the population $\mathcal{P}$ for a fixed time period T or under some given computational budget, that is, the maximum number of objective function(s) evaluations.

2.2 Limitations

The DE algorithm has gained remarkable popularity in the class of evolutionary algorithms and gathered a great amount of attention from the scientific community. The major limitations of the DE algorithm are premature convergence and stagnation. Gampler et al. [6], and Zaharie et al.[36] studied that the role of mutation strategy is instrumental for optimal performance in the DE algorithm. A mutation strategy sstartslosing its capacity to generate new vectors when vectors in the population accumulate around the basin of a local optimum, resulting in premature convergence and stagnation. Qin et al. [19] investigated that the performance of a mutation strategy is highly dependent on the landscape of the objective function, and utilization of a single mutation strategy is not sufficient for the optimal performance of the DE algorithm. Authors laid the foundations for utilizing multiple mutation strategies online by proposing a robust DE variant, called the SaDE algorithm.

A family of online adaptive DE variants has been emerged since the inception of the SaDE algorithm [2, 15, 28–30, 33, 34]. However, utilizing multiple mutation strategies online is very similar to the multi-arm bandit (MAB) framework with stochastic arms. A discussion on MAB-based formulation is presented in the next section.

3   Multi-arm Bandit Formulation for Utilizing Multiple Mutation Strategies

The Multi-Arm Bandits (MABs) models the situation where an agent or a decision maker faces the dilemma of optimal choice making in the presence of multiple choices. The rewards for choices are not known, beforehand, and reward for a choice is only realized after executing it. When rewards are static (but, unknown), the underlying MAB is called static, and when the rewards are stochastic, the underlying MAB is called dynamic. As, mutation strategies are stochastic in behavior, dynamic MAB framework is suitable for this study.

Suppose $\mathcal{M} = \{\mu_1, \mu_2, \ldots, \mu_n\}$ is the set of n different mutation strategies. The set $\mathcal{M}$ is the set of arms or choices for a bandit which is a DE variant. Say, $\mathcal{Q}^t = \{q_1^t, q_2^t, \ldots, q_n^t\}$ is the set of selection probabilities or quality values, such that, $\sum_{i=1}^{n} q_i^t = 1$, and $\mathcal{R}^t = \{r_1^t, r_2^t, \ldots, r_n^t\}$ is the reward set for each arm $\mu_i, i = 1, 2, \ldots, n$ at time stamp t. A bandit, or DE variant, selects an arm μ_i based on $\mathcal{Q}^t$ and receives a reward from $\mathcal{R}^t$. An agent is tasked with adjusting the quality values, also called q-values, so that the expected reward $\sum_{t=1}^{T} \mathbb{E}(\mathcal{R}^t)$ is maximized over the time period T.

Many bandit-based algorithms such as adaptive pursuit [26], probability matching [8], and $Q-$learning [31] are utilized for dealing with the above scenario. Undoubtedly, major DE variants with multiple mutation strategies are developed integrating these algorithms. A discussion on probability matching, adaptive pursuit, Upper Confidence Bound (UCB), and Q-learning algorithm is presented in the following subsections.

3.1   Probability Matching

In the probability matching algorithm [11], the "selection probability" q_k^t of each arm μ_k at time step t is updated using the equation

$$q_k^{t+1} = q_{min} + (1 - n \cdot q_{min}) \frac{\tilde{q}_k^{t+1}}{\sum_{k=1}^{n} \tilde{q}_k^{t+1}} \tag{3}$$

where $\tilde{q}_k^{t+1}$ is the empirical estimate of the kth arm, and $q_{min} \in (0, 1)$ is fixed minimum probability of each arm and it usually set between $0 < q_{min} < \frac{1}{n}$, where n is total number of arms. The empirical estimates are updated using the reward values so that the selection probability of the best arm $\mu_* = \arg\max[Q^t]$ converges to $q_{max} = q_{min} + (1 - n \cdot q_{min})$.

3.2   Adaptive Pursuit Algorithm

The adaptive pursuit algorithm proposed by Thierens [26] increases the selection probability of the best arm $\mu_* = \arg\max[\mathcal{Q}^t]$ at time step t using Eq. (4).

$$q_*^{t+1} = q_*^t + \alpha \cdot (q_{max} - q_*^t) \tag{4}$$

While, decreases the selection probabilities of other available arms ($\neq \mu_*$) using the Eq. (5).

$$q_k^{t+1} = q_k^t + \alpha(q_{min} - q_k^t) \quad \mu_k \neq \mu_*, k = 1, 2, \ldots, n - 1. \tag{5}$$

The parameter α in Eqs. (4) and (5) is called learning rate with values in the range $(0, 1]$. The q_{min} is a fixed minimum selection probability to the each arm so that the selection probability of an arm does not converge to zero. The value of q_{max} is set using the Eq. (6).

$$q_{max} = 1 - (n - 1) \cdot q_{min} \tag{6}$$

where n is the number of available arms.

3.3 Upper Confidence Bound (UCB) Algorithm

The upper confidence bound (UCB) algorithm [1] initially select each arm once using the Eq. (7).

$$N_k = 1, \qquad \hat{q}_k = r_k, \qquad \forall \mu_k \in \mathcal{M}, \tag{7}$$

where r_k denotes the reward obtained when an arm k is selected for the first time. For $t > K$, the empirical mean reward of action k is defined using the Eq. (8).

$$\hat{\mu}_k^t = \frac{1}{N_k^t} \sum_{\tau=1}^{t-1} r_k^\tau \, \mathbb{I}\{\mu^\tau = k\}, \tag{8}$$

where N_k^t is the selection frequency of the action μ_k up to time t, and $\mathbb{I}\{\cdot\}$ is the indicator function. At time $t > K$, an arm is selected using the following Eq. (9).

$$\mu_k^t = \arg\max_{k \in \mathcal{M}} \left(\hat{\mu}_k^t + \sqrt{\frac{2 \ln t}{N_k^t}} \right) \tag{9}$$

The first term promotes exploitation, while the second term is an exploration bonus that decreases as N_k^t increases. After observing reward r_k^t, the statistics N_k^t is updated using Eqs. (10) and (11).

$$N_k^{t+1} = N_k^t + \mathbb{I}\{\mu^t = k\}, \tag{10}$$

$$\hat{\mu}_k^{t+1} = \begin{cases} \dfrac{N_k^t \hat{\mu}_k^t + r_k^t}{N_t(k) + 1}, & \text{if } k = \mu_k, \\ \hat{\mu}_k^t, & \text{otherwise.} \end{cases} \tag{11}$$

It is important to note that all the above arm selection techniques, that is, the probability matching, the adaptive pursuit, and the UCB algorithm, do not utilize the explicit definition of the states of the environment. These algorithms belong to the class of reinforcement learning algorithms that deal with stateless sequential decision-making problems. The Q-learning [32] algorithm is an another class of reinforcement learning algorithm that utilizes the state information to generate a sequence of arm selection for maximizing the expected rewards. A discussion on the Q-learning algorithm is presented in the next subsection.

3.4 Q-Learning Algorithm

The Q-learning algorithm [31,32] proposed by Watkins compute the q-values of an action (or, arm) $\mu \in \mathcal{M}$ by associating it with the states $s \in \mathcal{S}$ of the environment $\mathcal{E}$. The q-values of state-action pairs is updated using the Eq. (12).

$$q^{t+1}(s_\ell, \mu_k) = q^t(s_\ell, \mu_k) + \alpha \left[r_k^t + \gamma \max_{\mu_{k'} \in \mathcal{M}} q^t(s_{\ell'}, \mu_{k'}) - q^t(s_\ell, \mu_k) \right] \quad (12)$$

where $s_{\ell,\ell'} \in \mathcal{S} = \{s_1, s_2, \ldots, s_m\}$ are states of the state space $\mathcal{S}$ and $\mu_{k,k'} \in \mathcal{M} = \{\mu_1, \mu_2, \ldots, \mu_n\}$ are actions of the action space $\mathcal{M}$. The α is the learning rate and γ is the discount rate with values in the range $(0, 1]$. The Q-learning algorithm maintains a state-action table of size $m \times n$ called Q-table, and it is also studied as a tabular RL algorithm.

This paper studies the existing DE variants utilizing multiple mutation strategy online based on the utilization of above-mentioned RL algorithms. A discussion on the reviewed DE variants is presented in the next section.

4 DE Variants with Multiple Mutation Strategies

4.1 Probability Matching Based DE Variants

Gong et al. [10,11] investigated the utility of probability matching in online mutation strategy selection in the DE algorithm. The authors proposed a DE variant called PM-AdapSS algorithm. Fialho et al. [5] utilizes the probability matching technique in the Differential Evolution algorithm for studying its performance in benchmark black-box optimization problems. Sharma et al. [21] proposed a DE variant called RecPM-DE that utilizes the recursive probability matching algorithm for mutation strategy in the Differential Evolution. DE variants such as JADE [37], CoDE [28], MPEDE [15], and CS-DE [17] indirectly employ the techniques of probability matching as they assume empirical estimates for selecting mutation strategies are sampled from a uniform probability distribution.

4.2 Adaptive Pursuit Based DE Variants

Gong et al. [10] utilized the adaptive pursuit algorithm to manage multiple mutation strategies of the DE algorithm for solving continuous optimization problems. Venske et al. [27] employed the probability matching algorithm for online selecting multiple mutation strategies to solve multi-objective optimization problems.

Zhang et al. [38] utilized the adaptive pursuit algorithm along with selective similarity rule to propose a DE variant called SCSS-L-SHADE algorithm. Bajpai et al. [2] integrated a modified adaptive pursuit technique to proposed a new DE variant called dmss-DE-pap algorithm.

4.3 Upper Confidence Bound (UCB)-Based DE Variants

Gonccalves et al. [9] utilized the upper confidence bound algorithm to manage multiple mutation strategies of the DE algorithm for solving multi-objective optimization problems. Meidani et al.[16] employed the upper confidence bound algorithm to propose hyperparameter optimizer called MAB-OS algorithm. Gao et al. [7] integrated the UCB algorithm for point generation in SaDE [19] algorithm and solved antenna design problem.

4.4 Q-Learning Based DE Variants

Kizilay et al.[13] incorporated the Q-learning algorithm to propose a DE variant called DE-QL for solving engineering design problems. Huynh et al. [12] utilized the Q-learning algorithm to propose qlDE algorithm for solving structural optimization problems. Tan et al. [25] integrated the Q-learning algorithm for online mutation strategy selection in the DE algorithm to solve numerical optimization problems. Bajpai and Bansal [3] employed the Q-learning algorithm to propose a new DE variant called the QLiDE algorithm for solving the 100-digit challenge [18]. Yu et al.[35] used the Q-learning algorithm for online selecting mutation strategies to solve a feature selection problem. Li et al. [14] utilized the Q-learning algorithm to propose a DE variant for solving dynamic job-scheduling problems.

It is important to note that these DE variants utilize Q-learning for online mutation strategy selection without using any a priori information regarding the underlying optimization problem. An advanced category of DE variants has emerged in the DE research that utilizes historical data of the objective function values and their relation with selected mutation strategies. Sharma et al. [20] integrated a trained artificial neural network along with the Q-learning for online mutation strategy selection in the DE algorithm. Tan et al.[24] integrated an artificial neural network along with Q-learning for online mutation strategy selection in the DE algorithm.

This paper reviewed the DE variants that utilize bandit-based reinforcement learning algorithms for strategy adaptation. To its best efforts, this paper includes a discussion on major DE variants published till the year 2025. The next section concludes this paper and presents future research directions.

5 Conclusion and Future Scope

Out of many branches of DE variants, this paper only focuses on the DE variants with multiple mutation strategies that utilize online RL techniques as their adaptation mechanism. The selection of mutation strategies online is a learning problem under uncertainty. Multi-arm bandits naturally fit into this setting and offer frameworks for developing improved DE variants.

RL-based online mutation strategy selection in the DE algorithm can be broadly studied as a sequential decision-making problem, where the choice of a mutation strategy influences both immediate fitness improvement and long-term search dynamics. The future scope in the direction of RL-based DE variants can be the exploration of richer state representations that describe population diversity, convergence behavior, and characteristics of the fitness landscape. Another possible investigation is to connect theoretical nuances of sequential decision-making techniques with the RL-based DE variants.

References

1. Auer, P., Cesa-Bianchi, N., Fischer, P.: Finite-time analysis of the multiarmed bandit problem. Mach. Learn. **47**(2), 235–256 (2002)
2. Bajpai, P., Anicho, O., Nagar, A.K., Bansal, J.C.: Dynamic mutation strategy selection in differential evolution using perturbed adaptive pursuit. SN Comput. Sci. **5**(6), 771 (2024)
3. Bajpai, P., Bansal, J.C.: Solving 100-digit challenge using q-learning inspired differential evolution algorithm. In: 2024 IEEE 3rd World Conference on Applied Intelligence and Computing (AIC), pp. 673–679. IEEE (2024)
4. Das, S., Mullick, S.S., Suganthan, P.N.: Recent advances in differential evolution–an updated survey. Swarm Evol. Comput. **27**, 1–30 (2016)
5. Fialho, Á., Schoenauer, M., Sebag, M.: Fitness-auc bandit adaptive strategy selection vs. the probability matching one within differential evolution: an empirical comparison on the bbob-2010 noiseless testbed. In: Proceedings of the 12th Annual Conference Companion on Genetic and Evolutionary Computation, pp. 1535–1542 (2010)
6. Gämperle, R., Müller, S.D., Koumoutsakos, P.: A parameter study for differential evolution. Adv. Intell. Syst. fuzzy Syst. Evol. Comput. **10**(10), 293–298 (2002)
7. Gao, T.Y., Jiao, Y.C., Zhang, Y.X., Zhang, L.: A hybrid self-adaptive differential evolution algorithm with simplified bayesian local optimizer for efficient design of antennas. IEEE Trans. Antennas Propag. (2024)
8. Goldberg, D.E.: Probability matching, the magnitude of reinforcement, and classifier system bidding. Mach. Learn. **5**, 407–425 (1990)
9. Gonçalves, R.A., Almeida, C.P., Pozo, A.: Upper confidence bound (ucb) algorithms for adaptive operator selection in moea/d. In: International Conference on Evolutionary Multi-Criterion Optimization, pp. 411–425. Springer (2015)
10. Gong, W., Fialho, Á., Cai, Z.: Adaptive strategy selection in differential evolution. In: Proceedings of the 12th Annual Conference on Genetic and Evolutionary Computation, pp. 409–416 (2010)
11. Gong, W., Fialho, Á., Cai, Z., Li, H.: Adaptive strategy selection in differential evolution for numerical optimization: an empirical study. Inf. Sci. **181**(24), 5364–5386 (2011)
12. Huynh, T.N., Do, D.T., Lee, J.: Q-learning-based parameter control in differential evolution for structural optimization. Appl. Soft Comput. **107**, 107464 (2021)

13. Kizilay, D., Tasgetiren, M.F., Oztop, H., Kandiller, L., Suganthan, P.N.: A differential evolution algorithm with q-learning for solving engineering design problems. In: 2020 IEEE Congress on Evolutionary Computation (CEC), pp. 1–8. IEEE (2020)

14. Li, X., et al.: A q-learning improved differential evolution algorithm for human-centric dynamic distributed flexible job shop scheduling problem. J. Manuf. Syst. **80**, 794–823 (2025)

15. Mallipeddi, R., Suganthan, P.N., Pan, Q.K., Tasgetiren, M.F.: Differential evolution algorithm with ensemble of parameters and mutation strategies. Appl. Soft Comput. **11**(2), 1679–1696 (2011)

16. Meidani, K., Mirjalili, S., Farimani, A.B.: Mab-os: multi-armed bandits metaheuristic optimizer selection. Appl. Soft Comput. **128**, 109452 (2022)

17. Meng, Z., Zhong, Y., Yang, C.: Cs-de: cooperative strategy based differential evolution with population diversity enhancement. Inf. Sci. **577**, 663–696 (2021)

18. Price, K., Awad, N.H., Ali, M.Z., Suganthan, P.: The 2019 100-digit challenge on real-parameter, single objective optimization: analysis of results. Department Computer Science, University of Freiburg, Freiburg, Germany, School of Computer Information Systems, Jordan University of Science and Technology, Jordan, and School of EEE, Nanyang Technol. Univ., Singapore, Rep, vol. 2019 (2019)

19. Qin, A.K., Huang, V.L., Suganthan, P.N.: Differential evolution algorithm with strategy adaptation for global numerical optimization. IEEE Trans. Evol. Comput. **13**(2), 398–417 (2008)

20. Sharma, M., Komninos, A., López-Ibáñez, M., Kazakov, D.: Deep reinforcement learning based parameter control in differential evolution. In: Proceedings of the Genetic and Evolutionary Computation Conference, pp. 709–717 (2019)

21. Sharma, M., López-Ibáñez, M., Kazakov, D.: Performance assessment of recursive probability matching for adaptive operator selection in differential evolution. In: International Conference on Parallel Problem Solving from Nature, pp. 321–333. Springer (2018)

22. Storn, R., Price, K.: Differential evolution-a simple and efficient heuristic for global optimization over continuous spaces. J. Global Optim. **11**, 341–359 (1997)

23. Sutton, R.S., Barto, A.G.: Reinforcement Learning: an Introduction. MIT Press (2018)

24. Tan, Z., Li, K., Wang, Y.: Differential evolution with adaptive mutation strategy based on fitness landscape analysis. Inf. Sci. **549**, 142–163 (2021)

25. Zhiping Tan, Yu., Tang, K.L., Huang, H., Luo, S.: Differential evolution with hybrid parameters and mutation strategies based on reinforcement learning. Swarm Evol. Comput. **75**, 101194 (2022)

26. Thierens, D.: An adaptive pursuit strategy for allocating operator probabilities. In: Proceedings of the 7th Annual Conference on Genetic and Evolutionary Computation, pp. 1539–1546 (2005)

27. Venske, S.M., Gonçalves, R.A., Delgado, M.R.: Ademo/d: Multiobjective optimization by an adaptive differential evolution algorithm. Neurocomputing **127**, 65–77 (2014)

28. Wang, J., Liao, J., Zhou, Y., Cai, Y.: Differential evolution enhanced with multiobjective sorting-based mutation operators. IEEE Trans. Cybern. **44**(12), 2792–2805 (2014)

29. Wang, Y., Cai, Z., Zhang, Q.: Differential evolution with composite trial vector generation strategies and control parameters. IEEE Trans. Evol. Comput. **15**(1), 55–66 (2011)

30. Wang, Y., Liu, Z., Wang, G.-G.: Improved differential evolution using two-stage mutation strategy for multimodal multi-objective optimization. Swarm Evol. Comput. **78**, 101232 (2023)
31. Watkins, C.J.C.H., Dayan, P.: Q-learning. Mach. Learn. **8**, 279–292 (1992)
32. Watkins, C.J.C.H.: Learning from delayed rewards. Phd Thesis (1989)
33. Wu, G., Mallipeddi, R., Suganthan, P.N., Wang, R., Chen, H.: Differential evolution with multi-population based ensemble of mutation strategies. Inf. Sci. **329**, 329–345 (2016)
34. Wu, G., Shen, X., Li, H., Chen, H., Lin, A., Suganthan, P.N.: Ensemble of differential evolution variants. Inf. Sci. **423**, 172–186 (2018)
35. Xiaobing, Yu., Zhengpeng, H., Luo, W., Xue, Yu.: Reinforcement learning-based multi-objective differential evolution algorithm for feature selection. Inf. Sci. **661**, 120185 (2024)
36. Zaharie, D.: Control of population diversity and adaptation in differential evolution algorithms. In Proc. MENDEL **9**, 41–46 (2003)
37. Zhang, J., Sanderson, A.C.: Jade: adaptive differential evolution with optional external archive. IEEE Trans. Evol. Comput. **13**(5), 945–958 (2009)
38. Zhang, S.X., Chan, W.S., Tang, K.S., Zheng, S.Y.: Adaptive strategy in differential evolution via explicit exploitation and exploration controls. Appl. Soft Comput. **107**, 107494 (2021)

Open Access This chapter is licensed under the terms of the Creative Commons Attribution-NonCommercial-NoDerivatives 4.0 International License (http:// creativecommons.org/licenses/by-nc-nd/4.0/), which permits any noncommercial use, sharing, distribution and reproduction in any medium or format, as long as you give appropriate credit to the original author(s) and the source, provide a link to the Creative Commons license and indicate if you modified the licensed material. You do not have permission under this license to share adapted material derived from this chapter or parts of it.

The images or other third party material in this chapter are included in the chapter's Creative Commons license, unless indicated otherwise in a credit line to the material. If material is not included in the chapter's Creative Commons license and your intended use is not permitted by statutory regulation or exceeds the permitted use, you will need to obtain permission directly from the copyright holder.

Author Index

© The Editor(s) (if applicable) and The Author(s) 2026
J. C. Bansal et al. (Eds.): SCIS 2025, LNNS 1929, pp. 203–204, 2026.
https://doi.org/10.1007/978-3-032-22911-3

MIX
Papier aus verantwortungsvollen Quellen
Paper from responsible sources
FSC® C105338

If you have any concerns about our products,
you can contact us on
ProductSafety@springernature.com

In case Publisher is established outside the EU,
the EU authorized representative is:
**Springer Nature Customer Service Center GmbH
Europaplatz 3, 69115 Heidelberg, Germany**

Printed by Libri Plureos GmbH
in Hamburg, Germany